ESSENTIALS OF MARINE BIOLOGY

Edited by
Nada Ben Abdallah

Arcler Press

Essentials of Marine Biology

Edited by Nada Ben Abdallah

Edition 2018

ArclerPress

Published by
Arcler Press LLC
708 3rd Avenue, 6th Floor
New York
NY 10017
United States of America
www.arclerpress.com

ISBN: 978-1-68094-540-9 (Hardback)

Printed in United States of America

Library of Congress Control Number: 2017933110

About the Editor

Nada Ben Abdallah

Nada graduated from the University of Zurich in 2010 and holds a PhD in neuroscience. Her research focuses on elucidating the processes of learning and memory as well as the affective processes and their underlying neural mechanisms. Her work so far involved normal and diseased animal models with a main interest in neurodegenerative diseases. She is currently working as a freelance researcher and science editor.

Table of Contents

3 MARINE MICROORGANISMS

4 MARINE INVERTEBRATES

5 MARINE FUNGI

6 MARINE VERTEBRATE

7 MARINE RESOURCES CONSERVATION

8 CORAL REEF

9 MARINE POLLUTION

List of Abbreviations

IUCN	International Union for Conservation of Nature
LMEs	Large Marine Ecosystems
LUCA	Last Universal Common Ancestor
MPAs	Marine Protected Areas
MTSG	Marine Turtle Specialist Group
PSAT or PAT	Pop-up Satellite Archival Tag
RFID	Radio-frequency Identification
TEDs	Turtle Excluder Devices

Preface

Basically, marine biology is the study of life in the oceans and other saltwater environments such as estuaries and wetlands. Marine biology is the scientific study of organisms in the ocean or other marine bodies of water. Given that in biology many phyla, families, and genera have some species that live in the sea and others that live on land, marine biology classifies species based on the environment rather than on taxonomy. Marine life is a vast resource, providing food, medicine, and raw materials, in addition to helping to support recreation and tourism all over the world. At a fundamental level, marine life helps determine the very nature of our planet. Marine organisms contribute significantly to the oxygen cycle, and are involved in the regulation of the Earth's climate. Shorelines are in part shaped and protected by marine life, and some marine organisms even help create new land. Many species are economically important to humans, including both finfish and shellfish. It is also becoming understood that the well-being of marine organisms and other organisms are linked in fundamental ways. The human body of knowledge regarding the relationship between life in the sea and important cycles is rapidly growing, with new discoveries being made nearly every day. These cycles include those of matter (such as the carbon cycle) and of air (such as Earth's respiration, and movement of energy through ecosystems including the ocean). Large areas beneath the ocean surface still remain effectively unexplored. As inhabitants of the largest environment on Earth, microbial marine systems drive changes in every global system. Microbes are responsible for virtually all the photosynthesis that occurs in the ocean, as well as the cycling of carbon, nitrogen, phosphorus and other nutrients and trace elements.

Content Coverage

Chapter One presents the history of marine biology that combines aspects of organismal function, ecological interaction and the study of marine biodiversity.

Chapters Two and *Three* cover the marine ecosystems for the overall health of marine environments and the diversity of microorganisms, respectively. A teaspoon of seawater contains about one million viruses. Most of these are bacteriophages, which are harmless to plants and animals, and are in fact essential to the regulation of saltwater and freshwater ecosystems. They infect and destroy bacteria in aquatic microbial communities, and are the most important mechanism of recycling carbon in the marine environment.

Chapter Four discusses about the marine invertebrates, marine invertebrates have a large variety of body plans, and have been categorised into over 30 phyla. They make up most of the macroscopic life in the oceans.

Chapter Five and *Six* present about the marine fungi and marine vertebrates, respectively. Marine fungi are species of fungi that live in marine environments and marine vertebrates include fish, seabirds, marine reptiles, and marine mammals.

Chapters Seven is marine conservation focuses on limiting human-caused damage to marine ecosystems, and on restoring damaged marine ecosystems. Marine conservation also focuses on preserving vulnerable marine species, and the history and nature of cytoplasm.

Chapter Eight provides the knowledge of coral reefs; diverse underwater ecosystems held together by calcium carbonate structures secreted by corals. Coral reefs are built by colonies of tiny animals found in marine waters that contain few nutrients.

Chapter Nine focuses on marine pollution that occurs when harmful, or potentially harmful, effects result from the entry into the ocean of chemicals, particles, industrial, agricultural and residential waste, noise, or the spread of invasive organisms.

CHAPTER 1

History of Marine Biology

Chapter Contents

The marine environment is the largest, most important, and most mysterious habitat on our planet. It contains more than 99 percent of the world's living space, produces half of its oxygen, plays a critical role in regulating its climate, and supports a remarkably diverse and exquisitely adapted array of life forms. As the 21st century progresses, human activities such as overfishing, coastal development, plastic pollution, oil spills, nutrient pollution, the spread of exotic species, and the emission of climate changing greenhouse gases are posing a significant threat to the marine environment and to many of its life forms.

INTRODUCTION

Marine biology is a hybrid subject that combines aspects of organismal function, ecological interaction and the study of marine biodiversity. The earliest studies of marine biology trace back to the Phoenicians and the Greeks who are known as the initial explorers of the oceans and their composition. The first recorded observations on the distribution and habits of marine life were made by Aristotle (384–322 BC).

Observations made in the earliest studies of marine biology provided an impetuous for the age of discovery and exploration that followed. During this time, a vast amount of knowledge was gained about life that exists in the oceans. Individuals who contributed significantly to this pool of knowledge include Captain James Cook (1728-1779), Charles Darwin (1809-1882) and Wyville Thomson(1830-1882).

The HMS *Challenger* during its pioneer expedition of 1872–76

These individuals took part in some of the more well-known expeditions of all time, making foundation contributions to marine biology. The era was important for the history of marine biology, but naturalists were still constrained by available technologies that limited their ability to effectively locate and accurately examine species that inhabited the deep parts of the ocean.

The subsequent creation of marine laboratories was another important development because marine scientists now had places to conduct research and process their specimens from expeditions. Technological advances, such as sound ranging, scuba diving gear, submersibles and remotely operated vehicles, progressively made it easier to study the deep ocean. This allowed marine biologists to explore depths people once thought never existed.

EARLY EXPLORATION OF THE DEEP SEA

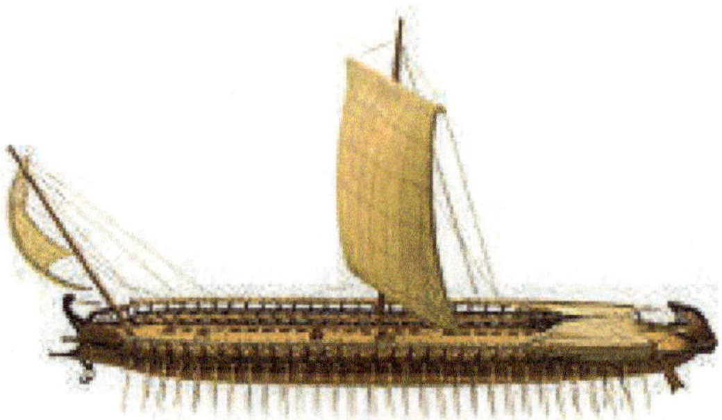

Model of a Greek boat

Phoenicians and Greeks

The history of marine biology can be traced as far back as 1200 BC when the Phoenicians and the Greeks began ocean voyages using celestial navigation. Phoenicians and Greeks were some of the first known explorers to leave their local communities bordering the Mediterranean Sea. They ventured outside the Mediterranean to the Atlantic Ocean with their knowledge of tides, currents and seasonal changes. It wasn't until much later at around 450 BC when observations of natural phenomena related to the oceans started getting recorded. Herodotus (484-425 BC) wrote of the regular tides in the Persian Gulf, the deposition of silt in the Nile Delta and used the term "Atlantic" to describe the western seas for the first time. It was during this time when many of the first recorded observations about the composition of the oceans.

Aristotle

Aristotle was one of the most notable Greek philosophers to examine and keep written records of observations concerning marine organisms. Aristotle identified a large range of species such as crustaceans, echinoderms, mollusks and fish. Aristotle has been known as the initiator of the tradition of natural philosophy and an influential player in the beginnings of marine biology because of the early observations he made about marine life

Hawaiian navigators sailing multi-hulled canoe

Polynesians

The Polynesians were also very involved in the exploration of marine life and their efforts are often overlooked.Throughout the time period of 300-1275 AD the Polynesians made efforts to explore and populate the great Polynesian triangle, which is bounded in the east by Easter Island, in the north by Hawaii and in the southwest by New Zealand. The Polynesians were among the first to go out and explore the mysteries of the ocean and marine life. In the years that followed the Polynesian efforts, there were minimal efforts that aimed to further man's understanding of the sea. This ended with the Age of Discovery in the late 15th century.

AGE OF DISCOVERY

Between the late 15th century and early 20th century, humans explored the oceans like never before creating new maps and charts and collecting specimens to bring back to their home ports. Most of the exploration that took place during this time was fueled by European countries such as Spain, Portugal, France, Italy, Scotland and Germany. Some of the landmark explorers of marine biology carried out their famous work during this time period. Explorers such as Captain James Cook, Charles Darwin and Wyville Thomson made revolutionary contributions to the history of marine biology during this time of exploration.

James Cook

James Cook is well known for his voyages of exploration for the British Navy in which he mapped out a significant amount of the world's uncharted waters. Cook's explorations took him around the world twice and led to countless descriptions of previously unknown plants and animals. Cook's explorations influenced many others and led to a number of scientists examining marine life more closely. Among those influenced was Charles Darwin who went on to make many contributions of his own.

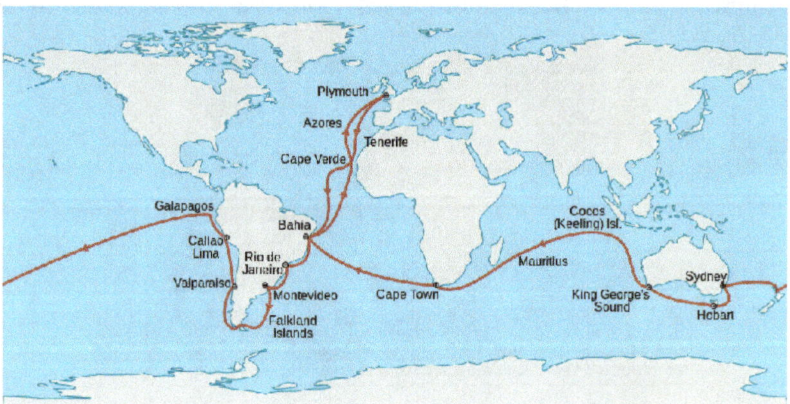

The voyage of the *Beagle*

Charles Darwin

Charles Darwin, best known for his theory of evolution, made many significant contributions to the early study of marine biology. He spent much of his time from 1831 to 1836 on the voyage of HMS *Beagle* collecting and studying specimens from a variety of marine organisms. It was also on this expedition where Darwin began to study coral reefs and their formation. He came up with the theory that the overall growth of corals is a balance between the growth of corals upward and the sinking of the sea floor. He then came up with the idea that wherever coral atolls would be found, the central island where the coral had started to grow would be gradually subsiding.

Charles Wyville Thomson

Another influential expedition was the voyage of HMS *Challenger* from 1872 to 1876, organized and later led by Charles Wyville Thomson. It was the first expedition purely devoted to marine science. The expedition collected and analyzed thousands of marine specimens, laying the foundation for present knowledge about life near the deep-sea floor. The findings from the expedition were a summary of the known natural, physical and chemical ocean science to that time.

Later Exploration

This era of marine exploration came to a close with the first and second round-the-world voyages of the Danish Galathea expeditions and Atlantic voyages by the USS *Albatross*, the first research vessel purpose built for marine research. These voyages further cleared the way for modern marine biology by building a base of knowledge about marine biology. This was followed by the progressive development of more advanced technologies which began to allow more extensive explorations of ocean depths that were once thought too deep to sustain life.

MODERN STUDIES

Marine Biology Laboratories

In the 1960s and 1970s, ecological research into the life of the ocean was undertaken at institutions set up specifically to study marine biology. Notable was the Woods Hole Oceanographic Institution in America, which established a model for other marine laboratories subsequently set up around the world. Their findings of unexpectedly high species diversity in places thought to be inhabitable stimulated much theorizing by population ecologists on how high diversification could be maintained in such a food-poor and seemingly hostile environment.

A science ROV being retrieved by an oceanographic research vessel

Exploration Technology

In the past, the study of marine biology has been limited by a lack of technology limiting the depth at which researchers could explore the ocean and the life within. Before the mid-twentieth century, the deep-sea bottom could not be seen unless one dredged a piece of it and brought it to the surface. This has changed dramatically due to the development of new technologies in both the laboratory and the open sea. These new technological developments have allowed scientists to explore parts of the ocean they did not even know existed.

The development of scuba gear allowed researchers to visually explore the oceans as it contains a self-contained underwater breathing apparatus allowing a person to breathe while being submerged 100 to 200 feet into the ocean. Submersibles were built like small submarines with the purpose of taking marine scientists to deeper depths of the ocean while protecting them from increasing atmospheric pressures that cause complications deep under water. The first models could hold several individuals and allowed limited visibility but enabled marine biologists to see and photograph the deeper portions of the oceans. Remotely operated underwater vehicles are now used with and without submersibles to see the deepest areas of the ocean that would be too dangerous for humans. ROVs are fully equipped with cameras and sampling equipment which allows researchers to see and control everything the vehicle does. ROVs have become the dominant type of technology used to view the deepest parts of the ocean.

Romanticization

In the late 20th century and into the 21st, marine biology was "glorified and romanticized through films and television shows," leading to an influx in interested students who required a damping on their enthusiasm with the day-to-day realities of the field.

REFERENCES

1. Levinton, Jeffrey S Marine biology: function, biodiversity, ecology. New York: Oxford University Press, 1995. p.3
2. History of the Study of Marine Biology MarineBio Conservation Society. Retrieved: April 7, 2014.
3. Levinton, Jeffrey S Marine biology: function, biodiversity, ecology. New York: Oxford University Press, 1995. p.4
4. Gage & Tyler (1991) p.1
5. Levinton, Jeffrey S. (1995) Marine biology: function, biodiversity, ecology, New York: Oxford University Press. p.7-8
6. A Brief History of Marine Biology and Oceanography. University of California. Retrieved 31 March 2014.
7. Ward (1974) p.5

8. Ward (1974) p.6
9. Levinton, Jeffrey S.. Marine biology: function, biodiversity, ecology. New York: Oxford University Press, 1995.
10. Levinton, Jeffrey S.. Marine biology: function, biodiversity, ecology. New York: Oxford University Press, 1995. p.5
11. Ward (1974) p.162
12. Gage & Tyler (1991) p.6
13. Rozwadowski, Helen M.. Fathoming the ocean the discovery and exploration of the deep sea. New ed. Cambridge, Mass.: Belknap, 2005. p173
14. Maienschein, Jane. 100 years exploring life, 1888-1988: the Marine Biological Laboratory at Woods Hole. Boston: Jones and Bartlett Publishers, 1989: 189-192
15. Gage & Tyler (1991) p.7
16. Ward (1974) p.142
17. Levinton, Jeffrey S.. Marine biology: function, biodiversity, ecology. New York: Oxford University Press, 1995. p.7
18. Anderson, Genny. Beginnings: History of Marine Science.
19. Woolston, Chris (11 December 2014). Marine Dreams (paper). Careers. Nature. 516 (7530): 277. doi:10.1038/nj7530-277a.

CHAPTER 2

Marine Ecosystem

Chapter Contents

- **Introduction**
- **Large Marine Ecosystem**
- **Marine Habitats**
- **Biome**

Marine ecology is the branch of ecological science concerned with organisms that live in or near the ocean, their behaviors, and their interactions with the environment. The scope of studies in marine ecology can range greatly, from examining unicellular microorganisms to researching global effects of pollution and human activity. Scientists might observe a specific population of organisms, identifying their behaviors and relationships, or investigate entire marine habitats to see how different living and nonliving factors contribute to the overall ecosystem.

INTRODUCTION

Marine ecosystems are among the largest of Earth`s aquatic ecosystems, salt marshes, intertidal zones, estuaries, lagoons, mangroves, coral reefs, the deep sea, and the sea floor. They can be contrasted with freshwater ecosystems, which have a lower salt content. Marine waters cover two-thirds of the surface of the Earth. Such places are considered ecosystems because the plant life supports the animal life and vice versa. See food chains.

Coral reefs form complex marine ecosystems with tremendous biodiversity.

Different types of starfish, coral reefs and fishes in the Great Barrier Reef

Marine ecosystems are very important for the overall health of both marine and terrestrial environments. According to the World Resource Center, coastal habitats alone account for approximately 1/3 of all marine biological productivity, and estuarine ecosystems (i.e. salt marshes, seagrasses, mangrove forests) are among the most productive regions on the planet. In addition, other marine ecosystems such as coral reefs, provide food and shelter to the highest levels of marine diversity in the world.

Marine ecosystems usually have a large biodiversity and are therefore thought to have a good resistance against invasive species. However, exceptions have been observed, and the mechanisms responsible in determining the success of an invasion are not yet clear.

LARGE MARINE ECOSYSTEMS

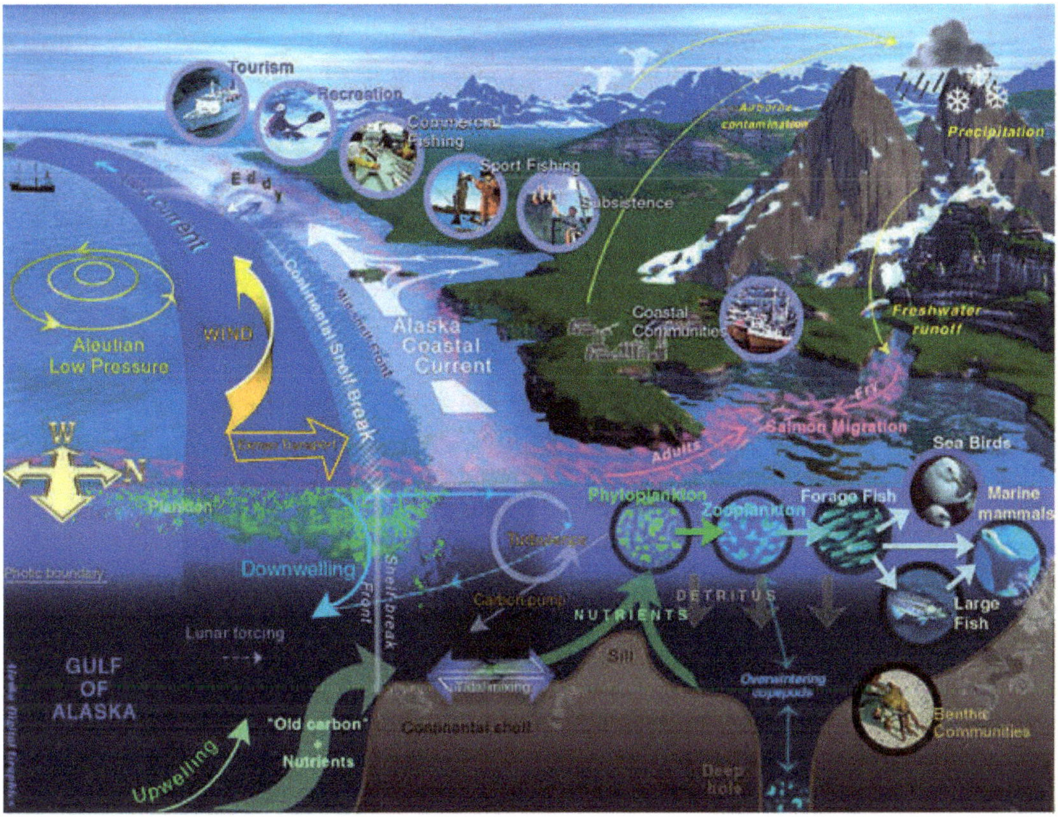

General characteristics of a large marine ecosystem (Gulf of Alaska)

Large marine ecosystems (LMEs) are regions of the world's oceans, encompassing coastal areas from river basins and estuaries to the seaward boundaries of continental shelves and the outer margins of the major ocean current systems. They are relatively large regions on the order of 200,000 km^2 or greater, characterized by distinct bathymetry, hydrography, productivity, and trophically dependent populations.

The system of LMEs has been developed by the US National Oceanic and Atmospheric Administration (NOAA) to identify areas of the oceans for conservation purposes. The objective is to use the LME concept as a tool for enabling ecosystem-based management to provide a collaborative approach to management of resources within ecologically-bounded transnational areas. This will be done in an international context and consistent with customary international law as reflected in 1982 UNConvention on the Law of the Sea.

LME-based conservation is based on recognition that the world's coastal ocean waters are degraded by unsustainable fishing practices, habitat degradation, eutrophication, toxic pollution, aerosol contamination, and emerging diseases, and that positive actions to mitigate these threats require coordinated actions by governments and civil society to recover depleted fish populations, restore degraded habitats and reduce coastal pollution.

Although the LMEs cover only the continental margins and not the deep oceans and oceanic islands, the 64 LMEs produce 95% of the world›s annual marine fishery biomass yields. Most of the global ocean pollution, overexploitation, and coastal habitat alteration occur within their waters. NOAA has conducted studies of principal driving forces affecting changes in biomass yields for 33 of the 64 LMEs, which have been peer-reviewed and published in ten volumes.

MARINE HABITATS

The marine environment supplies many kinds of habitats that support marine life. Marine life depends in some way on the saltwater that is in the sea (the term *marine* comes from the Latin *mare*, meaning sea or ocean). A habitat is an ecological or environmental area inhabited by one or more living species.

Marine habitats can be divided into coastal and open ocean habitats. Coastal habitats are found in the area that extends from as far as the tide comes in on the shoreline out to the edge of the continental shelf. Most marine life is found in coastal habitats, even though the shelf area occupies only seven percent of the total ocean area. Open ocean habitats are found in the deep ocean beyond the edge of the continental shelf.

Alternatively, marine habitats can be divided into pelagic and demersal zones. Pelagic habitats are found near the surface or in the open water column, away from the bottom of the ocean. Demersal habitats are near or on the bottom of the ocean. An organism living in a pelagic habitat is said to be a pelagic organism, as in pelagic fish. Similarly, an organism living in a demersal habitat is said to be a demersal organism, as in demersal fish. Pelagic habitats are intrinsically shifting and ephemeral, depending on what ocean currents are doing.

Marine habitats can be modified by their inhabitants. Some marine organisms, like corals, kelp, mangroves and seagrasses, are ecosystem engineers which reshape the marine environment to the point where they create further habitat for other organisms.

Only 29 percent of the world surface is land. The rest is ocean, home to the marine habitats. The oceans are nearly four kilometres deep on average and are fringed with coastlines that run for nearly 380,000 kilometres

In contrast to terrestrial habitats, marine habitats are shifting and ephemeral. Swimming organisms find areas by the edge of a continental shelf a good habitat, but only while upwellings bring nutrient rich water to the surface. Shellfish find habitat on sandy beaches, but storms, tides and currents mean their habitat continually reinvents itself.

The presence of seawater is common to all marine habitats. Beyond that many other things determine whether a marine area makes a good habitat and the type of habitat it makes. For example:

- Temperature – is affected by geographical latitude, ocean currents, weather, the discharge of rivers, and by the presence of hydrothermal vents or cold seeps.
- Sunlight – photosynthetic processes depend on how deep and turbid the water is.
- Nutrients – are transported by ocean currents to different marine habitats from land runoff, or by upwellings from the deep sea, or they sink though the sea as marine snow.
- Salinity – varies, particularly in estuaries or near river deltas, or by hydrothermal vents.
- Dissolved gases – oxygen levels in particular, can be increased by wave actions and decreased during algal blooms.
- Acidity – this is partly to do with dissolved gases above, since the acidity of the ocean is largely controlled by how much carbon dioxide is in the water.
- Turbulence – ocean waves, fast currents, and the agitation of water affect the nature of habitats.
- Cover – the availability of cover such as the adjacency of the sea bottom or the presence of floating objects.
- The occupying organisms themselves since organisms modify their habitats by the act of occupying them, and some, like corals, kelp, mangroves, and seagrasses create further habitats for other organisms.

Ocean	**Area million km²**	**%**	**Volume million cu km**	**%**	**Mean depth km**	**Max depth km**	**Coastline km**	**%**	**Ref**
Pacific Ocean	155.6	46.4	679.6	49.6	4.37	10.924	135,663		[4]
Atlantic Ocean	76.8	22.9	313.4	22.5	4.08	8.605	111,866		[5]
Indian Ocean	68.6	20.4	269.3	19.6	3.93	7.258	66,526		[4]
Southern Ocean	20.3	6.1	91.5	6.7	4.51	7.235	17,968		[6]
Arctic Ocean	14.1	4.2	17.0	1.2	1.21	4.665	45,389		[7]
Overall	**335.3**		**1370.8**		**4.09**	**10.924**	**377,412**		[8]

Land runoff, pouring into the sea, can contain nutrients

The ocean occupies 71% of the world surface, averaging nearly four kilometres in depth. There are five major oceans, of which the Pacific Ocean is nearly as large as the rest put together. Coastlines fringe the land for nearly 380,000 kilometres.

Marine habitats can be broadly divided into pelagic and demersal habitats. Pelagic habitats are the habitats of the open water column, away from the bottom of the ocean. Demersal habitats are the habitats that are near or on the bottom of the ocean. An organism living in a pelagic habitat is said to be a pelagic organism, as in pelagic fish. Similarly, an organism living in a demersal habitat is said to be a demersal organism, as in demersal fish. Pelagic habitats are intrinsically ephemeral, depending on what ocean currents are doing.

The land-based ecosystem depends on topsoil and fresh water, while the marine ecosystem depends on dissolved nutrients washed down from the land.

Ocean deoxygenation poses a threat to marine habitats, due to the growth of low oxygen zones.

Ocean Currents

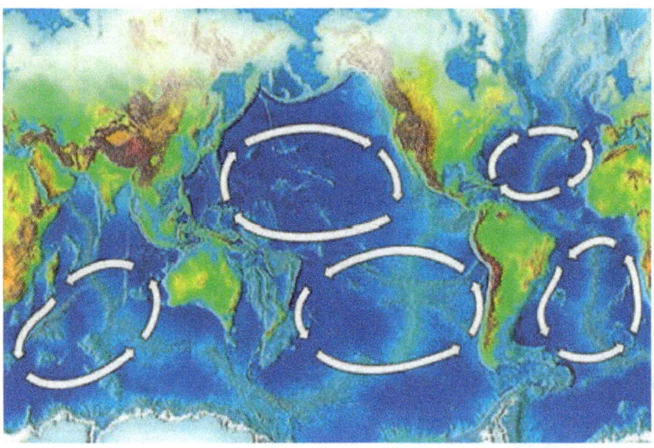

Ocean gyres rotate clockwise in the north and counterclockwise in the south

In marine systems, the ocean currents play a key role in determining which areas are effective as habitats, since they transport the basic nutrients needed to support marine life. Plankton are the life forms that inhabit the ocean and that are so small (less than 2 mm) that they cannot effectively propel themselves through the water, but must drift instead with the currents. If the current carries the right nutrients, and if it also flows at a suitably shallow depth where there is plenty of sunlight, then such a current itself can become a suitable habitat for photosynthesizing tiny algae called phytoplankton. These tiny plants are the primary producers in the ocean, at the start of the food chain. In turn, as the population of drifting phytoplankton grows, the water becomes a suitable habitat for zooplankton, which feed on the phytoplankton. While phytoplankton are tiny drifting plants, zooplankton are tiny drifting animals, such as the larvae of fish and marine invertebrates. If sufficient zooplankton establish themselves, the current becomes a candidate habitat for the forage fish that feed on them. Moreover if sufficient forage fish move to the area, it becomes a candidate habitat for larger predatory fish and other marine animals that feed on the forage fish. In this dynamic way, the current itself can, over time, become a moving habitat for multiple types of marine life.

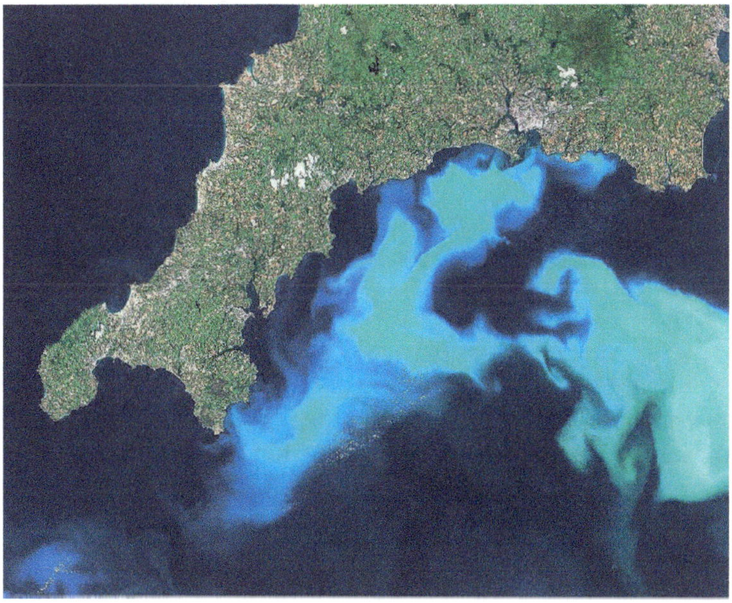

This algae bloom occupies sunlit epipelagic waters off the southern coast of England. The algae are maybe feeding on nutrients from land runoff or upwellings at the edge of the continental shelf

Ocean currents can be generated by differences in the density of the water. How dense water is depends on how saline or warm it is. If water contains differences in salt content or temperature, then the different densities will initiate a current. Water that is saltier or cooler will be denser, and will sink in relation to the surrounding water. Conversely, warmer and less salty water will float to the surface. Atmospheric winds and pressure differences also produces surface currents, waves and seiches. Ocean currents are also generated by the gravitational pull of the sun and moon (tides), and seismic activity (tsunami).

The rotation of the Earth affects the direction ocean currents take, and explains which way the large circular ocean gyres rotate in the image above left. Suppose a current at the equator is heading north. The Earth rotates eastward, so the water possesses that rotational momentum. But the further the water moves north, the slower the earth moves eastward. If the current could get to the North Pole, the earth would not be moving eastward at all. To conserve its rotational momentum, the further the current travels north the faster it must move eastward. So the effect is that the current curves to the right. This is the Coriolis effect. It is weakest at the equator and strongest at the poles. The effect is opposite south of the equator, where currents curve left.

Marine Topography

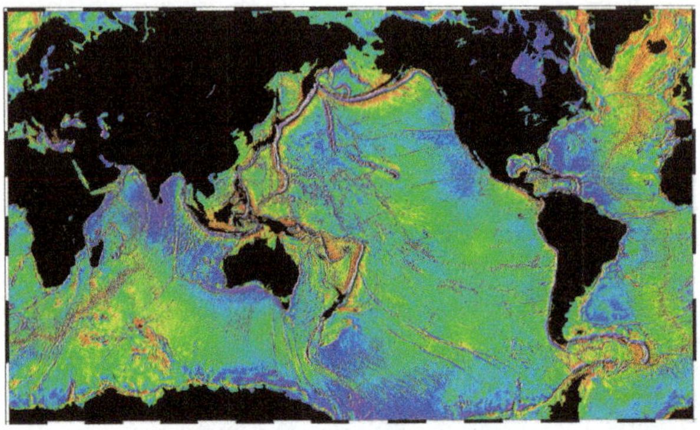

Map of underwater topography (1995, NOAA)

Marine topography refers to the shape the land has when it interfaces with the ocean. These shapes are obvious along coastlines, but they occur also in significant ways underwater. The effectiveness of marine habitats is partially defined by these shapes, including the way they interact with and shape ocean currents, and the way sunlight diminishes when these landforms occupy increasing depths.

Marine topographies include coastal and oceanic landforms ranging from coastal estuaries and shorelines to continental shelves and coral reefs. Further out in the open ocean, they include underwater and deep sea features such as ocean rises and seamounts. The submerged surface has mountainous features, including a globe-spanning mid-ocean ridge system, as well as undersea volcanoes, oceanic trenches, submarine canyons, oceanic plateaus and abyssal plains.

The mass of the oceans is approximately 1.35×10^{18} metric tons, or about 1/4400 of the total mass of the Earth. The oceans cover an area of 3.618×10^{8} km^2 with a mean depth of 3,682 m, resulting in an estimated volume of 1.332×10^{9} km^3.

Biomass

One measure of the relative importance of different marine habitats is the rate at which they produce biomass.

Producer	Biomass productivity ($gC/m^2/yr$)	Ref	Total area (million km^2)	Ref	Total production (billion tonnes C/yr)	Comment
swamps and marshes	2,500					Includes freshwater
coral reefs	2,000		0.28		0.56	
algal beds	2,000					

Producer	Biomass productivity (gC/m²/yr)	Ref	Total area (million km²)	Ref	Total production (billion tonnes C/yr)	Comment
river estuaries	1,800					
open ocean	125		311		39	

Coastal

Coastlines can be Volatile Habitats

Marine coasts are dynamic environments which constantly change, like the ocean which partially shape them. The Earth's natural processes, including weather and sea level change, result in the erosion, accretion and resculpturing of coasts as well as the flooding and creation of continental shelves and drowned river valleys.

The main agents responsible for deposition and erosion along coastlines are waves, tides and currents. The formation of coasts also depends on the nature of the rocks they are made of – the harder the rocks the less likely they are to erode, so variations in rock hardness result in coastlines with different shapes.

Tides often determine the range over which sediment is deposited or eroded. Areas with high tidal ranges allow waves to reach farther up the shore, and areas with lower tidal ranges produce deposition at a smaller elevation interval. The tidal range is influenced by the size and shape of the coastline. Tides do not typically cause erosion by themselves; however, tidal bores can erode as the waves surge up river estuaries from the ocean.

Waves erode coastline as they break on shore releasing their energy; the larger the wave the more energy it releases and the more sediment it moves. Sediment deposited by waves comes from eroded cliff faces and is moved along the coastline by the waves. Sediment deposited by rivers is the dominant influence on the amount of sediment located on a coastline.

Shores that look permanent through the short perceptive of a human lifetime are in fact among the most temporary of all marine structures.

The sedimentologist Francis Shepard classified coasts as *primary* or *secondary*.

- Primary coasts are shaped by non-marine processes, namely by changes in the land form. If a coast is in much the same condition as it was when sea level was stabilised after the last ice age, it is called a primary coast. "Primary coasts are created by erosion (the wearing away of soil or rock), deposition (the buildup of sediment or sand) or tectonic activity (changes in the structure of the rock and soil because of earthquakes). Many of these coastlines were formed as the sea level rose during the last 18,000 years, submerging river and glacial valleys to form bays and fjords." An example of a primary coast is a river delta, which forms when a river deposits soil and other material as it enters the sea.
- Secondary coasts are produced by marine processes, such as the action of the sea or by creatures that live in it. Secondary coastlines include sea cliffs, barrier islands, mud flats, coral reefs, mangrove swamps, and salt marshes.

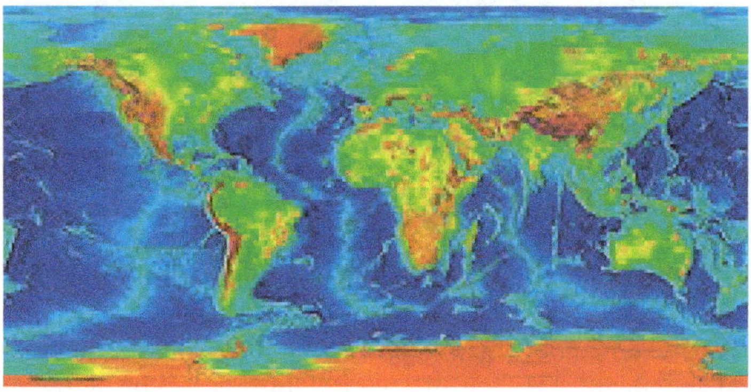

The global continental shelf, highlighted in cyan, defines the extent of coastal habitats, and occupies 5% of the total world area

Continental coastlines usually have a continental shelf, a shelf of relatively shallow water, less than 200 metres deep, which extends 68 km on average beyond the coast. Worldwide, continental shelves occupy a total area of about 24 million km^2 (9 million sq mi), 8% of the ocean›s total area and nearly 5% of the world›s total area. Since the continental shelf is usually less than 200 m deep, it follows that coastal habitats are generally photic, situated in the sunlit epipelagic zone. This means the conditions for photosynthetic processes so important for primary production, are available to coastal marine habitats. Because land is nearby, there are large discharges of nutrient rich land runoff into coastal waters. Further, periodic upwellings from the deep ocean can provide cool and nutrient rich currents along the edge of the continental shelf.

As a result, coastal marine life is the most abundant in the world. It is found in tidal pools, fjords and estuaries, near sandy shores and rocky coastlines, around coral reefs and on or above the continental shelf. Coastal fish include small forage fish as well as the larger predator fish that feed on them. Forage fish thrive in inshore waters where high productivity results from upwelling and shoreline run off of nutrients. Some are partial residents that spawn in streams, estuaries and bays, but most complete their life cycle in the zone. There can also be a mutualism between species that occupy adjacent marine habitats. For example, fringing reefs just below low tide level have a mutually beneficial relationship with mangrove forests at high tide level and sea grass meadows in between: the reefs protect the mangroves and seagrass from strong currents and waves that would damage them or erode the sediments in which they are rooted, while the mangroves and seagrass protect the coral from large influxes of silt, fresh water and pollutants. This additional level of variety in the environment is beneficial to many types of coral reef animals, which for example may feed in the sea grass and use the reefs for protection or breeding.

Coastal habitats are the most visible marine habitats, but they are not the only important marine habitats. Coastlines run for 380,000 km, and the total volume of the ocean is 1,370 million cubic km. This means that for each metre of coast, there is 3.6 cu km of ocean space available somewhere for marine habitats.

Waves and currents shape the intertidal shoreline, eroding the softer rocks and transporting and grading loose particles into shingles, sand or mud

Intertidal

Intertidal zones, those areas close to shore, are constantly being exposed and covered by the ocean's tides. A huge array of life lives within this zone.

Shore habitats range from the upper intertidal zones to the area where land vegetation takes prominence. It can be underwater anywhere from daily to very infrequently. Many species here are scavengers, living off of sea life that is washed up on the shore. Many land animals also make much use of the shore and intertidal habitats. A subgroup of organisms in this habitat bores and grinds exposed rock through the process of bioerosion.

Sandy Shores

Sandy shores provide shifting homes to many species

Sandy shores, also called beaches, are coastal shorelines where sand accumulates. Waves and currents shift the sand, continually building and eroding the shoreline. Longshore currents flow parallel to the beaches, making waves break obliquely on the sand. These currents transport large amounts of sand along coasts, forming spits, barrier islands and tombolos. Longshore currents also commonly create offshore bars, which give beaches some stability by reducing erosion.

Sandy shores are full of life, The grains of sand host diatoms, bacteria and other microscopic creatures. Some fish and turtles return to certain beaches and spawn eggs in the sand. Birds habitat beaches, like gulls, loons, sandpipers, terns and pelicans. Aquatic mammals, such sea lions, recuperate on them. Clams, periwinkles, crabs, shrimp, starfish and sea urchins are found on most beaches.

Sand is a sediment made from small grains or particles with diameters between about 60 μm and 2 mm. Mud (see mudflats below) is a sediment made from particles finer than sand. This small particle size means that mud particles tend to stick together, whereas sand particles do not. Mud is not easily shifted by waves and currents, and when it dries out, cakes into a solid. By contrast, sand is easily shifted by waves and currents, and when sand dries out it can be blown in the wind, accumulating into shifting sand dunes. Beyond the high tide mark, if the beach is low-lying, the wind can form rolling hills of sand dunes. Small dunes shift and reshape under the influence of the wind while larger dunes stabilise the sand with vegetation.

Ocean processes grade loose sediments to particle sizes other than sand, such as gravel or cobbles. Waves breaking on a beach can leave a berm, which is a raised ridge of coarser pebbles or sand, at the high tide mark. Shingle beaches are made of particles larger than sand, such as cobbles, or small stones. These beaches make poor habitats. Little life survives because the stones are churned and pounded together by waves and currents.

Rocky Shores

Tidepools on rocky shores make turbulent habitats for many forms of marine life

The relative solidity of rocky shores seems to give them a permanence compared to the shifting nature of sandy shores. This apparent stability is not real over even quite short geological time scales, but it is real enough over the short life of an organism. In contrast to sandy shores, plants and animals can anchor themselves to the rocks.

Competition can develop for the rocky spaces. For example, barnacles can compete successfully on open intertidal rock faces to the point where the rock surface is covered with them. Barnacles resist desiccation and grip well to exposed rock faces. However, in the crevices of the same rocks, the inhabitants are different. Here mussels can be the successful species, secured to the rock with their byssal threads.

Rocky and sandy coasts are vulnerable because humans find them attractive and want to live near them. An increasing proportion of the humans live by the coast, putting pressure on coastal habitats.

Mudflats

Mudflats become temporary habitats for migrating birds

Mudflats are coastal wetlands that form when mud is deposited by tides or rivers. They are found in sheltered areas such as bays, bayous, lagoons, and estuaries. Mudflats may be viewed geologically as exposed layers of bay mud, resulting from deposition of estuarine silts, clays and marine animal detritus. Most of the sediment within a mudflat is within the intertidal zone, and thus the flat is submerged and exposed approximately twice daily.

Mudflats are typically important regions for wildlife, supporting a large population, although levels of biodiversity are not particularly high. They are of particular importance to migratory birds. In the United Kingdom mudflats have been classified as a Biodiversity Action Plan priority habitat.

Mangrove and Salt Marshes

Mangroves provide nurseries for fish

Mangrove swamps and salt marshes form important coastal habitats in tropical and temperate areas respectively.

Mangroves are species of shrubs and medium size trees that grow in saline coastal sediment habitats in the tropics and subtropics – mainly between latitudes 25° N and 25° S. The saline conditions tolerated by various species range from brackish water, through pure seawater 30–40 parts per thousand (ppt), to water concentrated by evaporation to over twice the salinity of ocean seawater (up to 90 ppt). There are many mangrove species, not all closely related. The term mangrove is used generally to cover all of these species, and it can be used narrowly to cover just mangrove trees of the genus *Rhizophora*.

Mangroves form a distinct characteristic saline woodland or shrubland habitat, called a *mangrove swamp* or *mangrove forests*. Mangrove swamps are found in depositional coastal environments, where fine sediments (often with high organic content) collect in areas protected from high-energy wave action. Mangroves dominate three quarters of tropical coastlines.

Estuaries

Estuaries occur when rivers flow into a coastal bay or inlet. They are nutrient rich and have a transition zone which moves from freshwater to saltwater.

An estuary is a partly enclosed coastal body of water with one or more rivers or streams flowing into it, and with a free connection to the open sea. Estuaries form a transition zone between river environments and ocean environments and are subject to both marine influences, such as tides, waves, and the influx of saline water; and riverine influences, such as flows of fresh water and sediment. The inflow of both seawater and freshwater provide high levels of nutrients in both the water column and sediment, making estuaries among the most productive natural habitats in the world.

Most estuaries were formed by the flooding of river-eroded or glacially scoured valleys when sea level began to rise about 10,000-12,000 years ago. They are amongst the most heavily populated areas throughout the world, with about 60% of the world's population living along estuaries and the coast. As a result, estuaries are suffering degradation by many factors, including sedimentation from soil erosion from deforestation; overgrazing and other poor farming practices; overfishing; drainage and filling of wetlands; eutrophication due to excessive nutrients from sewage and animal wastes; pollutants including heavy metals, PCBs, radionuclides and hydrocarbons from sewage inputs; and diking or damming for flood control or water diversion.

Estuaries provide habitats for a large number of organisms and support very high productivity. Estuaries provide habitats for salmon and sea trout nurseries, as well as migratory bird populations. Two of the main characteristics of estuarine life are the variability in salinity and sedimentation. Many species of fish and invertebrates have various methods to control or conform to the shifts in salt concentrations and are termed osmoconformers and osmoregulators. Many animals also burrow to avoid predation and to live in the more stable sedimental environment. However, large numbers of bacteria are found within the sediment which have a very high oxygen demand. This reduces the levels of oxygen within the sediment often resulting in partially anoxic conditions, which can be further exacerbated by limited water flux. Phytoplankton are key primary producers in estuaries. They move with the water bodies and can be flushed in and out with the tides. Their productivity is largely dependent on the turbidity of the water. The main phytoplankton present are diatoms and dinoflagellates which are abundant in the sediment.

Kelp Forests

Kelp forests provide habitat for many marine organisms

Kelp forests are underwater areas with a high density of kelp. They form some of the most productive and dynamic ecosystems on Earth. Smaller areas of anchored kelp are called *kelp beds*. Kelp forests occur worldwide throughout temperate and polar coastal oceans.

Kelp forests provide a unique three-dimensional habitat for marine organisms and are a source for understanding many ecological processes. Over the last century, they have been the focus of extensive research, particularly in trophic ecology, and continue to provoke important ideas that are relevant beyond this unique ecosystem. For example, kelp forests can influence coastal oceanographic patterns and provide many ecosystem services.

However, humans have contributed to kelp forest degradation. Of particular concern are the effects of overfishing nearshore ecosystems, which can release herbivores from their normal population regulation and result in the over-grazing of kelp and other algae. This can rapidly result in transitions to barren landscapes where relatively few species persist.

Frequently considered an ecosystem engineer, kelp provides a physical substrate and habitat for kelp forest communities.In algae (Kingdom: Protista), the body of an individual organism is known as a thallus rather than as a plant (Kingdom: Plantae). The morphological structure of a kelp thallus is defined by three basic structural units:

- The holdfast is a root-like mass that anchors the thallus to the sea floor, though unlike true roots it is not responsible for absorbing and delivering nutrients to the rest of the thallus.
- The stipe is analogous to a plant stalk, extending vertically from the holdfast and providing a support framework for other morphological features.
- The fronds are leaf- or blade-like attachments extending from the stipe, sometimes along its full length, and are the sites of nutrient uptake and photosynthetic activity.

In addition, many kelp species have pneumatocysts, or gas-filled bladders, usually located at the base of fronds near the stipe. These structures provide the necessary buoyancy for kelp to maintain an upright position in the water column.

The environmental factors necessary for kelp to survive include hard substrate (usually rock), high nutrients (e.g., nitrogen, phosphorus), and light (minimum annual irradiance dose > 50 E m^{-2}). Especially productive kelp forests tend to be associated with areas of significant oceanographic upwelling, a process that delivers cool nutrient-rich water from depth to the ocean's mixed surface layer. Water flow and turbulence facilitate nutrient assimilation across kelp fronds throughout the water column. Water clarity affects the depth to which sufficient light can be transmitted. In ideal conditions, giant kelp (*Macrocystis spp.*) can grow as much as 30-60 centimetres vertically per day. Some species such as *Nereocystis* are annual while others like *Eisenia* are perennial, living for more than 20 years. In perennial kelp forests, maximum growth rates occur during upwelling months (typically spring and summer) and die-backs correspond to reduced nutrient availability, shorter photoperiods and increased storm frequency.

Seagrass Meadows

White-spotted puffers like living in seagrass areas

Seagrasses are flowering plants from one of four plant families which grow in marine environments. They are called *seagrasses* because the leaves are long and narrow and are very often green, and because the plants often grow in large meadows which look like grassland. Since seagrasses photosynthesize and are submerged, they must grow submerged in the photic zone, where there is enough sunlight. For this reason, most occur in shallow and sheltered coastal waters anchored in sand or mud bottoms.

Seagrasses form extensive beds or meadows, which can be either monospecific (made up of one species) or multispecific (where more than one species co-exist). Seagrass beds make highly diverse and productive ecosystems. They are home to phyla such as juvenile and adult fish, epiphytic and free-living macroalgae and microalgae, mollusks, bristle worms, and nematodes. Few species were originally considered to feed directly on seagrass leaves (partly because of their low nutritional content), but scientific reviews and improved working methods have shown that seagrass herbivory is a highly important link in the food chain, with hundreds of species feeding on seagrasses worldwide, including green turtles, dugongs, manatees, fish, geese, swans, sea urchins and crabs.

Seagrasses are ecosystem engineers in the sense that they partly create their own habitat. The leaves slow down water-currents increasing sedimentation, and the seagrass roots and rhizomes stabilize the seabed. Their importance to associated species is mainly due to provision of shelter (through their three-dimensional structure in the water column), and due to their extraordinarily high rate of primary production. As a result, seagrasses provide coastal zones with ecosystem services, such as fishing grounds, wave protection, oxygen production and protection against coastal erosion. Seagrass meadows account for 15% of the ocean's total carbon storage.

Coral Reefs

Reefs comprise some of the densest and most diverse habitats in the world. The best-known types of reefs are tropical coral reefs which exist in most tropical waters; however, reefs can also exist in cold water. Reefs are built up by corals and other calcium-depositing animals, usually on top of a rocky outcrop on the ocean floor. Reefs can also grow on other surfaces, which has made it possible to create artificial reefs. Coral reefs also support a huge community of life, including the corals themselves, their symbiotic zooxanthellae, tropical fish and many other organisms.

Much attention in marine biology is focused on coral reefs and the El Niño weather phenomenon. In 1998, coral reefs experienced the most severe mass bleaching events on record, when vast expanses of reefs across the world died because sea surface temperatures rose well above normal. Some reefs are recovering, but scientists say that between 50% and 70% of the world's coral reefs are now endangered and predict that global warming could exacerbate this trend.

Open Ocean

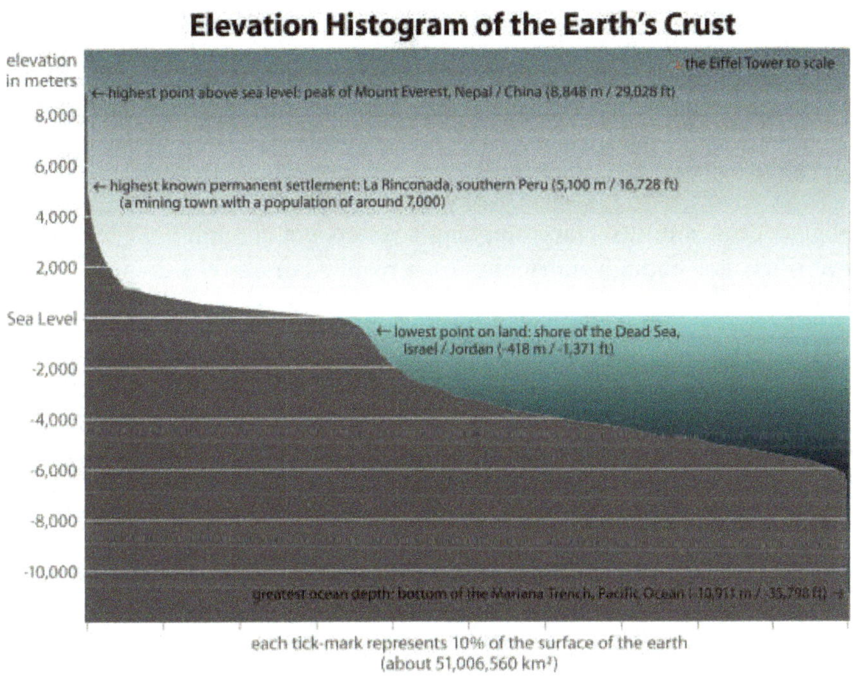

Elevation-area graph showing the proportion of land area at given heights and the proportion of ocean area at given depths

The open ocean is relatively unproductive because of a lack of nutrients, yet because it is so vast, it has more overall primary production than any other marine habitat. Only about 10 % of marine species live in the open ocean. But among them are the largest and fastest of all marine animals, as well as the animals that dive the deepest and migrate the longest. In the depths lurk animal that, to our eyes, appear hugely alien.

Surface Waters

In the open ocean, sunlit surface epipelagic waters get enough light for photosynthesis, but there are often not enough nutrients. As a result, large areas contain little life apart from migrating animals.

The surface waters are sunlit. The waters down to about 200 m are said to be in the epipelagic zone. Enough sunlight enters the epipelagic zone to allow photosynthesis by phytoplankton. The epipelagic zone is usually low in nutrients. This partially because the organic debris produced in the zone, such as excrement and dead animals, sink to the depths and are lost to the upper zone. Photosynthesis can happen only if both sunlight and nutrients are present.

In some places, like at the edge of continental shelves, nutrients can upwell from the ocean depth, or land runoff can be distributed by storms and ocean currents. In these areas, given that both sunlight and nutrients are now present, phytoplankton can rapidly establish itself, multiplying so fast that the water turns green from the chlorophyll, resulting in an algal bloom. These nutrient rich surface waters are among the most biologically productive in the world, supporting billions of tonnes of biomass.

- Phytoplankton are eaten by zooplankton–small animals that, like phytoplankton, drift in the ocean currents. The most abundant zooplankton species are copepods and krill: tiny crustaceans that are the most numerous animals on earth. Other types of zooplankton include jelly fish and the larvae of fish, marine worms, starfish, and other marine organisms. In turn, the zooplankton are eaten by filter-feeding animals, including some seabirds, small forage fish like herrings and sardines, whale sharks, manta rays, and the largest animal in the world, the blue whale. Yet again, moving up the foodchain, the small forage fish are in turn eaten by larger predators, such as tuna, marlin, sharks, large squid, seabirds, dolphins, and toothed whales.

Deep sea

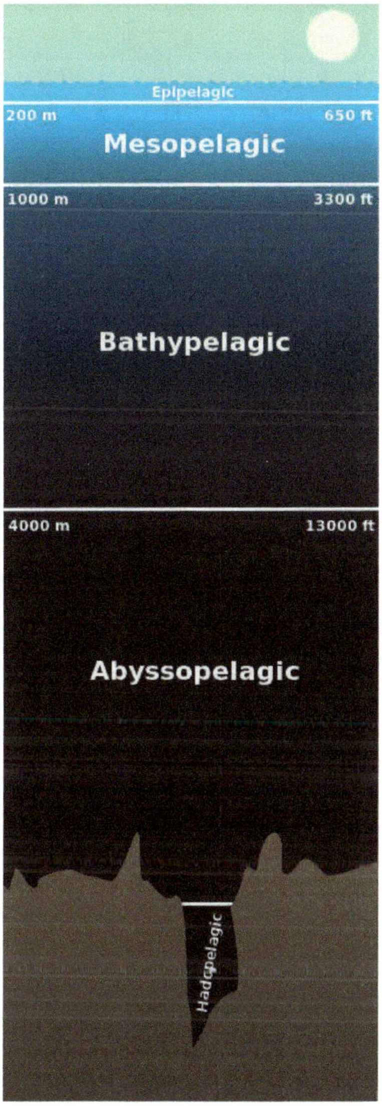

Scale diagram of the layers of the pelagic zone

The deep sea starts at the aphotic zone, the point where sunlight loses most of its energy in the water. Many life forms that live at these depths have the ability to create their own light a unique evolution known as bio-luminescence.

In the deep ocean, the waters extend far below the epipelagic zone, and support very different types of pelagic life forms adapted to living in these deeper zones.

Much of the aphotic zone›s energy is supplied by the open ocean in the form of detritus. In deep water, marine snow is a continuous shower of mostly organic detritus falling from the upper layers of the water column. Its origin lies in activities within the productive photic zone. Marine snow includes dead or dying plankton, protists (diatoms), fecal matter, sand, soot and other inorganic dust. The snowflakes grow over time and may reach several centimetres in diameter, travelling for weeks before reaching the ocean floor. However, most organic components of marine snow are consumed by microbes, zooplankton and other filter-feeding animals within the first 1,000 metres of their journey, that is, within the epipelagic zone. In this way marine snow may be considered the foundation of deep-sea mesopelagic and benthic ecosystems: As sunlight cannot reach them, deep-sea organisms rely heavily on marine snow as an energy source.

Some deep-sea pelagic groups, such as the lanternfish, ridgehead, marine hatchetfish, and lightfish families are sometimes termed *pseudoceanic* because, rather than having an even distribution in open water, they occur in significantly higher abundances around structural oases, notably seamounts and over continental slopes. The phenomenon is explained by the likewise abundance of prey species which are also attracted to the structures.

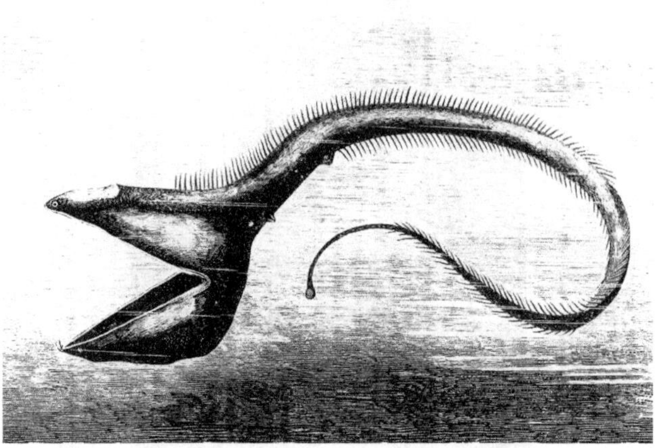

The umbrella mouth gulper eel can swallow a fish much larger than itself

The fish in the different pelagic and deep water benthic zones are physically structured, and behave in ways, that differ markedly from each other. Groups of coexisting species within each zone all seem to operate in similar ways, such as the small mesopelagic vertically migrating plankton-feeders, the bathypelagic anglerfishes, and the deep water benthic rattails.

Ray finned species, with spiny fins, are rare among deep sea fishes, which suggests that deep sea fish are ancient and so well adapted to their environment that invasions by more modern fishes have been unsuccessful. The few ray fins that do exist are mainly in the Beryciformes and Lampriformes, which are also ancient forms. Most deep sea pelagic fishes belong to their own orders, suggesting a long evolution in deep sea environments. In contrast, deep water benthic species, are in orders that include many related shallow water fishes.

The umbrella mouth gulper is a deep sea eel with an enormous loosely hinged mouth. It can open its mouth wide enough to swallow a fish much larger than itself, and then expand its stomach to accommodate its catch.

Sea floor

Vents and seeps

Zooarium chimney provides a habitat for vent biota

Hydrothermal vents along the mid-ocean ridge spreading centers act as oases, as do their opposites, cold seeps. Such places support unique biomes and many new microbes and other lifeforms have been discovered at these locations.

Trenches

The deepest recorded oceanic trenches measure to date is the Mariana Trench, near the Philippines, in the Pacific Ocean at 10,924 m (35,838 ft). At such depths, water pressure is extreme and there is no sunlight, but some life still exists. A white flatfish, a shrimp and a jellyfish were seen by the American crew of the bathyscaphe *Trieste* when it dove to the bottom in 1960.

Seamounts

Marine life also flourishes around seamounts that rise from the depths, where fish and other sea life congregate to spawn and feed.

BIOME

A **biome** is a formation of plants and animals that have common characteristics due to similar climates and can be found over a range of continents. Spanning continents, Biomes are distinct biological communities that have formed in response to a shared physical climate. Biomes are distinct from habitats, because any biome can comprise a variety of habitats. A biome contrasts with a microbiome. A microbiome is also a mix of organisms that coexist in a defined space, but on a much smaller scale. For example, the human microbiome is the collection of bacteria, viruses, and other microorganisms that are present on a human.

History of the Concept

The term was suggested in 1916 by Clements, originally as a synonym for biotic community of Möbius (1877). Later, it gained its current definition, based on earlier concepts of phytophysiognomy, formation and vegetation (used in opposition to flora), with the inclusion of the animal element and the exclusion of the taxonomic element of species composition. In 1935, Tansley would add the climatic and soil aspects to the idea, calling it ecosystem. The International Biological Program (1964-74) projects would popularize the concept of biome.

However, in some contexts, the term biome is used in a different manner. In German literature, particularly in the Walter terminology, the term is used similarly as biotope (a concrete geographycal unit), while the biome definition used in this article is used as a international, non-regional, terminology - irrespectively of the continent in which an area is present, it takes the same biome name - and corresponds to his zonobiome, orobiome and pedobiome (biomes determinated by climate zone, altitude or soil).

In Brazilian literature, the term biome is sometimes used as synonym of biogeographic province, an area based on species composition (the term floristic province being used when plant species are considered), or also as synonym of the morphoclimatic and phytogeographical domain of Apostrophe, a geographic space with subcontinental dimensions, with the predominance of similar geomorphologic and climatic characteristics, and of a certain vegetation form. Indeed, both include many biomes.

Classifications

Biomes are defined by climate regimes and biogeography.

A 1978 study on North American grasslands found a positive logistic correlation between evapotranspiration in mm/yr and above-ground net primary production in $g/m^2/yr$. The general results from the study were that precipitation and water use led to above-ground primary production, while solar irradiation and temperature lead to below-ground primary production (roots), and temperature and water lead to cool and warm season growth habit. These findings help explain the categories used in Holdridge's bioclassification scheme (see below), which were then later simplified by Whittaker. The number of classification schemes and the variety of determinants used in those schemes, however, should be taken as strong indicators that biomes do not fit perfectly into the classification schemes created.

Holdridge (1947, 1964) life Zones

Holdridge classified climates based on the biological effects of temperature and rainfall on vegetation under the assumption that these two abiotic factors are the largest determinants of the types of vegetation found in a habitat. Holdridge uses the four axes to define 30 so-called humidity provinces, which are clearly visible in his diagram. While this scheme largely ignores soil and sun exposure, Holdridge acknowledged that these were important.

Allee (1949) Biome-Types

The principal biome-types by Allee (1949):

- Tundra
- Taiga
- Deciduous forest
- Grasslands
- Desert
- High plateaus
- Tropical forest
- Minor terrestrial biomes

Kendeigh (1961) Biomes

The principal biomes of the world by Kendeigh (1961):

- Terrestrial
    - Temperate deciduous forest
    - Coniferous forest
    - Woodland
    - Chaparral
    - Tundra
    - Grassland
    - Desert
    - Tropical savanna
    - Tropical forest
- Marine
    - Oceanic plankton and nekton
    - Balanoid-gastropod-thallophyte
    - Pelecypod-annelid
    - Coral reef

Whittaker (1962, 1970, 1975) Biome-Types

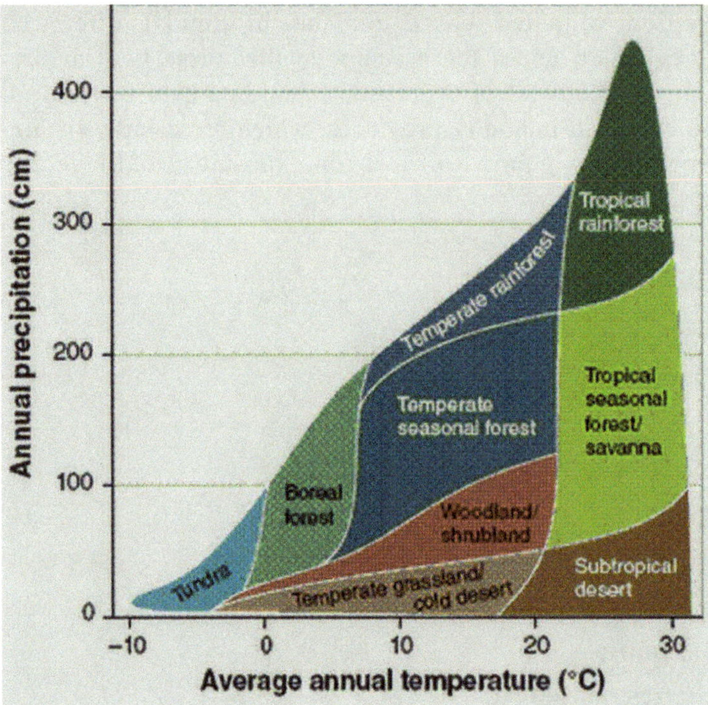

The distribution of vegetation types as a function of mean annual temperature and precipitation.

Whittaker classified biomes using two abiotic factors: precipitation and temperature. His scheme can be seen as a simplification of Holdridge`s; more readily accessible, but missing Holdridge`s greater specificity.

Whittaker based his approach on theoretical assertions and empirical sampling. He was in a unique position to make such a holistic assertion because he had previously compiled a review of biome classifications.

Key definitions for Understanding Whittaker's scheme

- Physiognomy: the apparent characteristics, outward features, or appearance of ecological communities or species.
- Biome: a grouping of terrestrial ecosystems on a given continent that are similar in vegetation structure, physiognomy, features of the environment and characteristics of their animal communities.
- Formation: a major kind of community of plants on a given continent.
- Biome-type: grouping of convergent biomes or formations of different continents, defined by physiognomy.
- Formation-type: a grouping of convergent formations.

Whittaker's distinction between biome and formation can be simplified: formation is used when applied to plant communities only, while biome is used when concerned with both plants and animals. Whittaker's convention of biome-type or formation-type is simply a broader method to categorize similar communities.

Whittaker`s Parameters for Classifying Biome-Types

Whittaker, seeing the need for a simpler way to express the relationship of community structure to the environment, used what he called "gradient analysis" of ecocline patterns to relate communities to climate on a worldwide scale. Whittaker considered four main ecoclines in the terrestrial realm.

- Intertidal levels: The wetness gradient of areas that are exposed to alternating water and dryness with intensities that vary by location from high to low tide
- Climatic moisture gradient
- Temperature gradient by altitude
- Temperature gradient by latitude

Along these gradients, Whittaker noted several trends that allowed him to qualitatively establish biome-types:

- The gradient runs from favorable to extreme, with corresponding changes in productivity.
- Changes in physiognomic complexity vary with how favorable of an environment exists (decreasing community structure and reduction of stratal differentiation as the environment becomes less favorable).
- Trends in diversity of structure follow trends in species diversity; alpha and beta species diversities decrease from favorable to extreme environments.
- Each growth-form (i.e. grasses, shrubs, etc.) has its characteristic place of maximum importance along the ecoclines.
- The same growth forms may be dominant in similar environments in widely different parts of the world.

Whittaker summed the effects of gradients (3) and (4) to get an overall temperature gradient, and combined this with gradient (2), the moisture gradient, to express the above conclusions in what is known as the Whittaker classification scheme. The scheme graphs average annual precipitation (x-axis) versus average annual temperature (y-axis) to classify biome-types.

Biome-Types

1. Tropical rainforest
2. Tropical seasonal rainforest
    - deciduous
    - semideciduous
3. Temperate giant rainforest

4. Montane rainforest
5. Temperate deciduous forest
6. Temperate evergreen forest
   - needleleaf
   - sclerophyll
7. Subarctic-subalpin needle-leaved forests (taiga)
8. Elfin woodland
9. Thorn forests and woodlands
10. Thorn scrub
11. Temperate woodland
12. Temperate shrublands
    - deciduous
    - heath
    - sclerophyll
    - subalpine-needleleaf
    - subalpine-broadleaf
13. Savanna
14. Temperate grassland
15. Alpine grasslands
16. Tundra
17. Tropical desert
18. Warm-temperate desert
19. Cool temperate desert scrub
20. Arctic-alpine desert
21. Bog
22. Tropical fresh-water swamp forest
23. Temperate fresh-water swamp forest
24. Mangrove swamp
25. Salt marsh
26. Wetland

Goodall (1974–) Ecosystem Types

The multiauthored series *Ecosystems of the world*, edited by David W. Goodall, provides a comprehensive coverage of the major "ecosystem types or biomes" on earth:

I. Terrestrial Ecosystems
   - A. Natural Terrestrial Ecosystems

- 1. Wet Coastal Ecosystems
- 2. Dry Coastal Ecosystems
- 3. Polar and Alpine Tundra
- 4. Mires: Swamp, Bog, Fen, and Moor
- 5. Temperate Deserts and Semi-Deserts
- 6. Coniferous Forests
- 7. Temperate Deciduous Forests
- 8. Natural Grasslands
- 9. Heathlands and Related Shrublands
- 10. Temperate Broad-Leaved Evergreen Forests
- 11. Mediterranean-Type Shrublands
- 12. Hot Deserts and Arid Shrublands
- 13. Tropical Savannas
- 14. Tropical Rain Forest Ecosystems
- 15. Wetland Forests
- 16. Ecosystems of Disturbed Ground

- B. Managed Terrestrial Ecosystems
  - 17. Managed Grasslands
  - 18. Field Crop Ecosystems
  - 19. Tree Crop Ecosystems
  - 20. Greenhouse Ecosystems
  - 21. Bioindustrial Ecosystems

II. Aquatic Ecosystems

- A. Inland Aquatic Ecosystems
  - 22. River and Stream Ecosystems
  - 23. Lakes and Reservoirs
- B. Marine Ecosystems

    - 24. Intertidal and Littoral Ecosystems
    - 25. Coral Reefs
    - 26. Estuaries and Enclosed Seas
    - 27. Ecosystems of the Continental Shelves
    - 28. Ecosystems of the Deep Ocean
  - C. Managed Aquatic Ecosystems
    - 29. Managed Aquatic Ecosystems

III. Underground Ecosystems

  - 30. Cave Ecosystems

Walter (1976, 2002) Zonobiomes

The eponymously-named Heinrich Walter classification scheme considers the seasonality of temperature and precipitation. The system, also assessing precipitation and temperature, finds nine major biome types, with the important climate traits and vegetation types. The boundaries of each biome correlate to the conditions of moisture and cold stress that are strong determinants of plant form, and therefore the vegetation that defines the region. Extreme conditions, such as flooding in a swamp, can create different kinds of communities within the same biome.

Zonobiome	Zonal soil type	Zonal vegetation type
ZB I. Equatorial, always moist, little temperature seasonality	Equatorial brown clays	Evergreen tropical rainforest
ZB II. Tropical, summer rainy season and cooler "winter" dry season	Red clays or red earths	Tropical seasonal forest, seasonal dry forest, scrub, or savanna
ZB III. Subtropical, highly seasonal, arid climate	Serosemes, sierozemes	Desert vegetation with considerable exposed surface
ZB IV. Mediterranean, winter rainy season and summer drought	Mediterranean brown earths	Sclerophyllous (drought-adapted), frost-sensitive shrublands and woodlands
ZB V. Warm temperate, occasional frost, often with summer rainfall maximum	Yellow or red forest soils, slightly podsolic soils	Temperate evergreen forest, somewhat frost-sensitive
ZB VI. Nemoral, moderate climate with winter freezing	Forest brown earths and grey forest soils	Frost-resistant, deciduous, temperate forest
ZB VII. Continental, arid, with warm or hot summers and cold winters	Chernozems to serozems	Grasslands and temperate deserts

ZB VIII. Boreal, cold temperate with cool summers and long winters	Podsols	Evergreen, frost-hardy, needle-leaved forest (taiga)
ZB IX. Polar, short, cool summers and long, cold winters	Tundra humus soils with solifluction (permafrost soils)	Low, evergreen vegetation, without trees, growing over permanently frozen soils

Schültz (1988) ecozones

Schültz (1988) defined nine ecozones (note that his concept of ecozone is more similar to the concept of biome used in this article than to the concept of ecozone of BBC):

1. polar/subpolar zone
2. boreal zone
3. humid mid-latitudes
4. arid mid-latitudes
5. tropical/subtropical arid lands
6. Mediterranean-type subtropics
7. seasonal tropics
8. humid subtropics
9. humid tropics

Bailey (1989) ecoregions

Robert G. Bailey nearly developed a biogeographical classification system of ecoregions for the United States in a map published in 1976. He subsequently expanded the system to include the rest of North America in 1981, and the world in 1989. The Bailey system, based on climate, is divided into seven domains (polar, humid temperate, dry, humid, and humid tropical), with further divisions based on other climate characteristics (subarctic, warm temperate, hot temperate, and subtropical; marine and continental; lowland and mountain).

- 100 Polar Domain
    - 120 Tundra Division (Köppen: Ft)
    - M120 Tundra Division – Mountain Provinces
    - 130 Subarctic Division (Köppen: E)
    - M130 Subarctic Division – Mountain Provinces
- 200 Humid Temperate Domain
    - 210 Warm Continental Division (Köppen: portion of Dcb)
    - M210 Warm Continental Division – Mountain Provinces

- 220 Hot Continental Division (Köppen: portion of Dca)
- M220 Hot Continental Division – Mountain Provinces
- 230 Subtropical Division (Köppen: portion of Cf)
- M230 Subtropical Division – Mountain Provinces
- 240 Marine Division (Köppen: Do)
- M240 Marine Division – Mountain Provinces
- 250 Prairie Division (Köppen: arid portions of Cf, Dca, Dcb)
- 260 Mediterranean Division (Köppen: Cs)
- M260 Mediterranean Division – Mountain Provinces

- 300 Dry Domain
  - 310 Tropical/Subtropical Steppe Division
  - M310 Tropical/Subtropical Steppe Division – Mountain Provinces
  - 320 Tropical/Subtropical Desert Division
  - 330 Temperate Steppe Division
  - 340 Temperate Desert Division
- 400 Humid Tropical Domain
  - 410 Savanna Division
  - 420 Rainforest Division

Olson & Dinerstein (1998) Biomes for WWF / Global 200

A team of biologists convened by the World Wildlife Fund (WWF) developed a scheme that divided the world's land area into biogeographic realms (called "ecozones in a BBC scheme), and these into ecoregions (Olson & Dinerstein, 1998–). Each ecoregion is characterized by a main biome (also called major habitat type).

This classification is used to define the Global 200 list of ecoregions identified by the WWF as priorities for conservation.

For the terrestrial ecoregions, there is a specific EcoID, format XXnnNN (XX is the biogeographic realm, nn is the biome number, NN is the individual number).

Biogeographic Realms (Terrestrial and Freshwater)

- NA: Nearctic

- PA: Palearctic
- AT: Afrotropic
- IM: Indomalaya
- AA: Australasia
- NT: Neotropic
- OC: Oceania
- AN: Antarctic

It should be noted, however, that the applicability of the realms scheme above - based on Udvardy (1975) - to most freshwater taxa is unresolved.

Biogeographic Realms (Marine)

- Arctic
- Temperate Northern Atlantic
- Temperate Northern Pacific
- Tropical Atlantic
- Western Indo-Pacific
- Central Indo-Pacific
- Eastern Indo-Pacific
- Tropical Eastern Pacific
- Temperate South America
- Temperate Southern Africa
- Temperate Australasia
- Southern Ocean

Biomes (terrestrial)

- 01: Tropical and subtropical moist broadleaf forests (tropical and subtropical, humid)
- 02: Tropical and subtropical dry broadleaf forests (tropical and subtropical, semihumid)
- 03: Tropical and subtropical coniferous forests (tropical and subtropical, semihumid)
- 04: Temperate broadleaf and mixed forests (temperate, humid)
- 05: Temperate coniferous forests (temperate, humid to semihumid)
- 06: Boreal forests/taiga (subarctic, humid)
- 07: Tropical and subtropical grasslands, savannas, and shrublands (tropical and subtropical, semiarid)
- 08: Temperate grasslands, savannas, and shrublands (temperate, semiarid)

- 09: Flooded grasslands and savannas (temperate to tropical, fresh or brackish water inundated)
- 10: Montane grasslands and shrublands (alpine or montane climate)
- 11: Tundra (Arctic)
- 12: Mediterranean forests, woodlands, and scrub or sclerophyll forests (temperate warm, semihumid to semiarid with winter rainfall)
- 13: Deserts and xeric shrublands (temperate to tropical, arid)
- 14: Mangrove (subtropical and tropical, salt water inundated)

Biomes (freshwater)

According to the WWF, the following are classified as freshwater biomes:

- Large lakes
- Large river deltas
- Polar freshwaters
- Montane freshwaters
- Temperate coastal rivers
- Temperate floodplain rivers and wetlands
- Temperate upland rivers
- Tropical and subtropical coastal rivers
- Tropical and subtropical floodplain rivers and wetlands
- Tropical and subtropical upland rivers
- Xeric freshwaters and endorheic basins
- Oceanic islands
- Streams and rivers

Biomes (marine)

Biomes of the coastal and continental shelf areas (neritic zone):

- Polar
- Temperate shelves and sea
- Temperate upwelling
- Tropical upwelling
- Tropical coral

Summary of the scheme

- Biosphere
- Biogeographic realms (terrestrial) (8)

- Ecoregions (867), each characterized by a main biome type (14)
- Ecosystems (biotopes)
- Biosphere
- Biogeographic realms (freshwater) (8)
- Ecoregions (426), each characterized by a main biome type (12)
- Ecosystems (biotopes)
- Biosphere
- Biogeographic realms (marine) (13)
- (Marine provinces) (62)
- Ecoregions (232), each characterized by a main biome type (5)
- Ecosystems (biotopes)

Example:

- Biosphere
- Biogeographic realm: Palearctic
- Ecoregion: Dinaric Mountains mixed forests (PA0418); biome type: temperate broadleaf and mixed forests
- Ecosystem: Orjen, vegetation belt between 1,100–1,450 m, Oromediterranean zone, nemoral zone (temperate zone)
- Biotope: *Oreoherzogio-Abietetum illyricae* Fuk. (Plant list)
- Plant: Silver fir (*Abies alba*)

Other Biomes

Marine Biomes

Pruvot (1896) zones or "systems":

- Litoral zone
- Pelagic zone
- Abyssal zone

Longhurst (1998) biomes:

- Coastal
- Polar
- Trade wind
- Westerly

Other marine habitat types (not covered yet by the Global 200/WWF scheme):

- Open sea
- Deep sea
- Hydrothermal vents
- Cold seeps
- Benthic zone
- Pelagic zone (trades and westerlies)
- Abyssal
- Hadal (ocean trench)
- Littoral/Intertidal zone
- Kelp forest
- Pack ice

Anthropogenic Biomes

Humans have altered global patterns of biodiversity and ecosystem processes. As a result, vegetation forms predicted by conventional biome systems can no longer be observed across much of Earth's land surface as they have been replaced by crop and rangelands or cities. Anthropogenic biomes provide an alternative view of the terrestrial biosphere based on global patterns of sustained direct human interaction with ecosystems, including agriculture, human settlements, urbanization, forestry and other uses of land. Anthropogenic biomes offer a new way forward in ecology and conservation by recognizing the irreversible coupling of human and ecological systems at global scales and moving us toward an understanding of how best to live in and manage our biosphere and the anthropogenic biomes we live in.

Major anthropogenic biomes:

- Dense settlements
- Croplands
- Rangelands
- Forested
- Indoor

Microbial biomes

Endolithic biomes

The endolithic biome, consisting entirely of microscopic life in rock pores and cracks, kilometers beneath the surface, has only recently been discovered, and does not fit well into most classification schemes.

Dermal Biome

The dermal biome is the living ecosystem that animals (including humans) have evolved, that permits them to live symbiotically and in balance with the microbes on and in them (the microbiome). This ecosystem consists of skin, follicles, hair, sebaceous glands, sweat glands, arrector pili muscles, peptides, proteins, lipids and its associated microbiota. A healthy dermal biome has several functions: it resists infection of pathogens, protects against moisture loss and water damage, dynamically regulates body temperature and supports the healthy renewal of skin through the epidermal cell life cycle.

- Infection Resistance: Commensal microbiota assist the dermal biome resist infection for pathogenic bacteria by i] out-competing pathogens for resources, ii] training or stimulating the host's immune system to defeat the pathogen, or iii] expressing substances that are directly hostile to the pathogen.
- Water-barrier regulation: The dermal biome regulates the water barrier preventing moisture from escaping (except when expressed as sweat) and preventing environmental water from permeating the skin. Because environmental water can have a chilling effect to mammals and warm-blooded animals when kept in close proximity to the epidermis, the dermal biome also produces hydrophobic lipids that repel water.
- Temperature regulation: The dermal biome is responsible for thermoregulation. To regulate excess heat, the dermal biome activates the sweat glands, allowing for evaporative cooling as sweat evaporates. To regulate cooler temperatures, the arrector pili muscles contract, causing hairs to "stand up" (goosebumps), and thereby trap an insulating blanket of air close to the skin.
- Skin renewal: a healthy biome supports the replacement of skin through the life cycle of epidermal cells as they proliferate in the basal layers of the epidermis until they die and are shed (desquamation).

An **aquatic ecosystem** is an ecosystem in a body of water. Communities of organisms that are dependent on each other and on their environment live in aquatic ecosystems. The two main types of aquatic ecosystems are marine ecosystems and freshwater ecosystems.

Types

Marine

Marine ecosystems cover approximately 71% of the Earth's surface and contain approximately 97% of the planet's water. They generate 32% of the world's net primary production. They are distinguished from freshwater ecosystems by the presence of dissolved compounds, especially salts, in the water. Approximately 85% of the dissolved materials in seawater are sodium and chlorine. Seawater has an average salinity of 35 parts per thousand (ppt) of water. Actual salinity varies among different marine ecosystems.

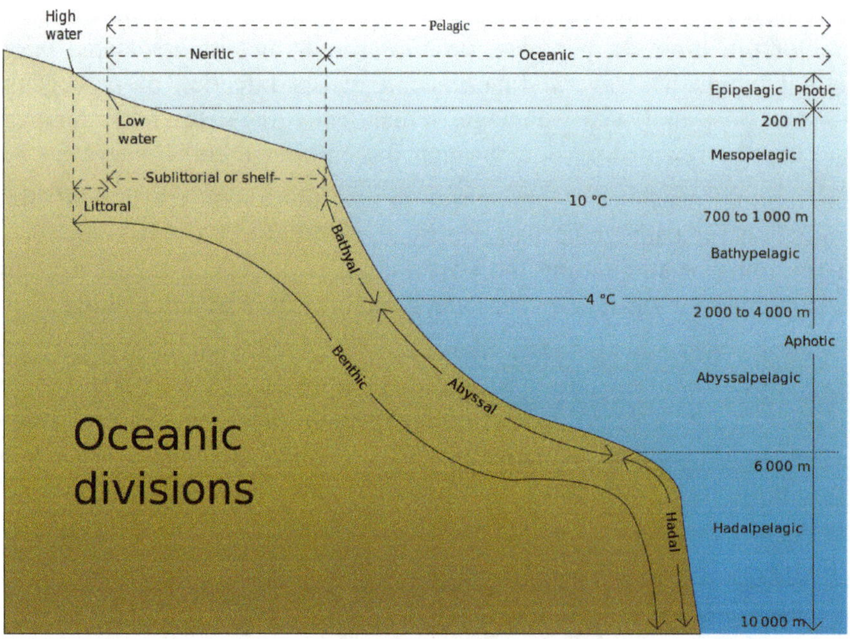

A classification of marine habitats.

Marine ecosystems can be divided into many zones depending upon water depth and shoreline features. The oceanic zone is the vast open part of the ocean where animals such as whales, sharks, and tuna live. The benthic zone consists of substrates below water where many invertebrates live. The intertidal zone is the area between high and low tides; in this figure it is termed the littoral zone. Other near-shore (neritic) zones can include estuaries, salt marshes, coral reefs, lagoons and mangrove swamps. In the deep water, hydrothermal vents may occur where chemosynthetic sulfur bacteria form the base of the food web.

Classes of organisms found in marine ecosystems include brown algae, dinoflagellates, corals, cephalopods, echinoderms, and sharks. Fishes caught in marine ecosystems are the biggest source of commercial foods obtained from wild populations.

Environmental problems concerning marine ecosystems include unsustainable exploitation of marine resources (e.g. overfishing of certain species), marine pollution, climate change, and building on coastal areas.

Freshwater

Freshwater ecosystem.

Freshwater ecosystems cover 0.78% of the Earth's surface and inhabit 0.009% of its total water. They generate nearly 3% of its net primary production. Freshwater ecosystems contain 41% of the world's known fish species.

There are three basic types of freshwater ecosystems:

- Lentic: slow moving water, including pools, ponds, and lakes.
- Lotic: faster moving water, for example streams and rivers.
- Wetlands: areas where the soil is saturated or inundated for at least part of the time.

Lentic

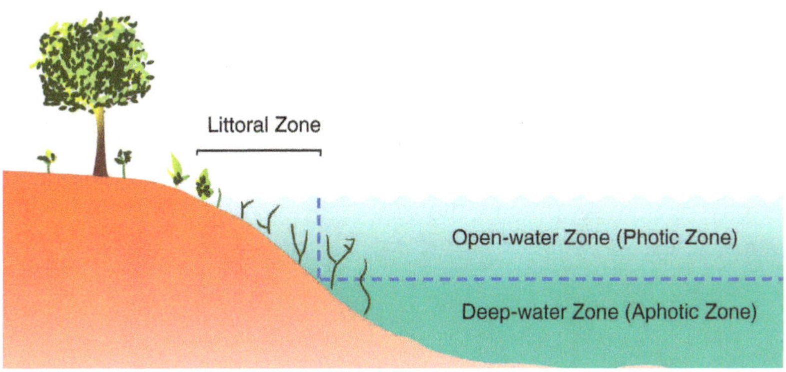

The three primary zones of a lake.

Lake ecosystems can be divided into zones. One common system divides lakes into three zones (see figure). The first, the littoral zone, is the shallow zone near the shore. This is where rooted wetland plants occur. The offshore is divided into two further zones, an open water zone and a deep water zone. In the open water zone (or photic zone) sunlight supports photosynthetic algae, and the species that feed upon them. In the deep water zone, sunlight is not available and the food web is based on detritus entering from the littoral and photic zones. Some systems use other names. The off shore areas may be called the pelagic zone, the photic zone may be called the limnetic zone and the aphotic zone may be called the profundal zone. Inland from the littoral zone one can also frequently identify a riparian zone which has plants still affected by the presence of the lake—this can include effects from windfalls, spring flooding, and winter ice damage. The production of the lake as a whole is the result of production from plants growing in the littoral zone, combined with production from plankton growing in the open water.

Wetlands can be part of the lentic system, as they form naturally along most lake shores, the width of the wetland and littoral zone being dependent upon the slope of the shoreline and the amount of natural change in water levels, within and among years. Often dead trees accumulate in this zone, either from windfalls on the shore or logs transported to the site during floods. This woody debris provides important habitat for fish and nesting birds, as well as protecting shorelines from erosion.

Two important subclasses of lakes are ponds, which typically are small lakes that intergrade with wetlands, and water reservoirs. Over long periods of time, lakes, or bays within them, may gradually become enriched by nutrients and slowly fill in with organic sediments, a process called succession. When humans use the watershed, the volumes of sediment entering the lake can accelerate this process. The addition of sediments and nutrients to a lake is known as eutrophication.

Ponds

Ponds are small bodies of freshwater with shallow and still water, marsh, and aquatic plants. They can be further divided into four zones: vegetation zone, open water, bottom mud and surface film. The size and depth of ponds often varies greatly with the time of year; many ponds are produced by spring flooding from rivers. Food webs are based both on free-floating algae and upon aquatic plants. There is usually a diverse array of aquatic life, with a few examples including algae, snails, fish, beetles, water bugs, frogs, turtles, otters and muskrats. Top predators may include large fish, herons, or alligators. Since fish are a major predator upon amphibian larvae, ponds that dry up each year, thereby killing resident fish, provide important refugia for amphibian breeding. Ponds that dry up completely each year are often known as vernal pools. Some ponds are produced by animal activity, including alligator holes and beaver ponds, and these add important diversity to landscapes.

Lotic

The major zones in river ecosystems are determined by the river bed's gradient or by the velocity of the current. Faster moving turbulent water typically contains greater concentrations of dissolved oxygen, which supports greater biodiversity than the slow moving water of pools. These distinctions form the basis for the division of rivers into upland and lowland rivers. The food base of streams within riparian forests is mostly derived from the trees, but wider streams and those that lack a canopy derive the majority of their food base from algae. Anadromous fish are also an important source of nutrients. Environmental threats to rivers include loss of water, dams, chemical pollution and introduced species. A dam produces negative effects that continue down the watershed. The most important negative effects are the reduction of spring flooding, which damages wetlands, and the retention of sediment, which leads to loss of deltaic wetlands.

Wetlands

Wetlands are dominated by vascular plants that have adapted to saturated soil. There are four main types of wetlands: swamp, marsh, fen and bog (both fens and bogs are types of mire). Wetlands are the most productive natural ecosystems in the world because of the proximity of water and soil. Hence they support large numbers of plant and animal species. Due to their productivity, wetlands are often converted into dry land with dykes and drains and used for agricultural purposes. The construction of dykes, and dams, has negative consequences for individual wetlands and entire watersheds. Their closeness to lakes and rivers means that they are often developed for human settlement. Once settlements are constructed and protected by dykes, the settlements then become vulnerable to land subsidence and ever increasing risk of flooding. The Louisiana coast around New Orleans is a well-known example; the Danube Delta in Europe is another.

Functions

Aquatic ecosystems perform many important environmental functions. For example, they recycle nutrients, purify water, attenuate floods, recharge ground water and provide habitats for wildlife. Aquatic ecosystems are also used for human recreation, and are very important to the tourism industry, especially in coastal regions.

The health of an aquatic ecosystem is degraded when the ecosystem›s ability to absorb a stress has been exceeded. A stress on an aquatic ecosystem can be a result of physical, chemical or biological alterations of the environment. Physical alterations include changes in water temperature, water flow and light availability. Chemical alterations include changes in the loading rates of biostimulatory nutrients, oxygen consuming materials, and toxins. Biological alterations include over-harvesting of commercial species and the introduction of exotic species. Human populations can impose excessive stresses on aquatic ecosystems. There are many examples of excessive stresses with negative consequences. Consider three. The environmental history of the Great Lakes of North America illustrates this problem, particularly how multiple stresses, such as water pollution, over-harvesting and invasive species can combine. The Norfolk Broadlands in England illustrate similar decline with pollution and invasive species. Lake Pontchartrain along the Gulf of Mexico illustrates the negative effects of different stresses including levee construction, logging of swamps, invasive species and salt water intrusion.

Abiotic Characteristics

An ecosystem is composed of biotic communities that are structured by biological interactions and abiotic environmental factors. Some of the important abiotic environmental factors of aquatic ecosystems include substrate type, water depth, nutrient levels, temperature, salinity, and flow. It is often difficult to determine the relative importance of these factors without rather large experiments. There may be complicated feedback loops. For example, sediment may determine the presence of aquatic plants, but aquatic plants may also trap sediment, and add to the sediment through peat.

The amount of dissolved oxygen in a water body is frequently the key substance in determining the extent and kinds of organic life in the water body. Fish need dissolved oxygen to survive, although their tolerance to low oxygen varies among species; in extreme cases of low oxygen some fish even resort to air gulping. Plants often have to produce aerenchyma, while the shape and size of leaves may also be altered. Conversely, oxygen is fatal to many kinds of anaerobic bacteria.

Nutrient levels are important in controlling the abundance of many species of algae. The relative abundance of nitrogen and phosphorus can in effect determine which species of algae come to dominate. Algae are a very important source of food for aquatic life, but at the same time, if they become over-abundant, they can cause declines in fish when they decay. Similar over-abundance of algae in coastal environments such as the Gulf of Mexico produces, upon decay, a hypoxic region of water known as a dead zone.

The salinity of the water body is also a determining factor in the kinds of species found in the water body. Organisms in marine ecosystems tolerate salinity, while many freshwater organisms are intolerant of salt. The degree of salinity in an estuary or delta is an important control upon the type of wetland (fresh, intermediate, or brackish), and the associated animal species. Dams built upstream may reduce spring flooding, and reduce sediment accretion, and may therefore lead to saltwater intrusion in coastal wetlands.

Freshwater used for irrigation purposes often absorbs levels of salt that are harmful to freshwater organisms.

Biotic Characteristics

The biotic characteristics are mainly determined by the organisms that occur. For example, wetland plants may produce dense canopies that cover large areas of sediment—or snails or geese may graze the vegetation leaving large mud flats. Aquatic environments have relatively low oxygen levels, forcing adaptation by the organisms found there. For example, many wetland plants must produce aerenchyma to carry oxygen to roots. Other biotic characteristics are more subtle and difficult to measure, such as the relative importance of competition, mutualism or predation. There are a growing number of cases where predation by coastal herbivores including snails, geese and mammals appears to be a dominant biotic factor.

Autotrophic Organisms

Autotrophic organisms are producers that generate organic compounds from inorganic material. Algae use solar energy to generate biomass from carbon dioxide and are possibly the most important autotrophic organisms in aquatic environments.Of course, the more shallow the water, the greater the biomass contribution from rooted and floating vascular plants. These two sources combine to produce the extraordinary production of estuaries and wetlands, as this autotrophic biomass is converted into fish, birds, amphibians and other aquatic species.

Chemosynthetic bacteria are found in benthic marine ecosystems. These organisms are able to feed on hydrogen sulfide in water that comes from volcanic vents. Great concentrations of animals that feed on these bacteria are found around volcanic vents. For example, there are giant tube worms (*Riftia pachyptila*) 1.5 m in length and clams (*Calyptogena magnifica*) 30 cm long.

Heterotrophic Organisms

Heterotrophic organisms consume autotrophic organisms and use the organic compounds in their bodies as energy sources and as raw materials to create their own biomass. Euryhaline organisms are salt tolerant and can survive in marine ecosystems, while stenohaline or salt intolerant species can only live in freshwater environments.

REFERENCES

1. Dickinson, C.I. 1963. British Seaweeds. The Kew Series
2. Abercrombie, M., Hickman, C.J. and Johnson, M.L. 1966.A Dictionary of Biology. Penguin Reference Books, London
3. The World`s Oceans and Seas. Encarta. Retrieved 19 April 2008.
4. CIA Factbook: Pacific ocean.
5. CIA Factbook: Atlantic ocean.
6. CIA Factbook: Indian ocean.
7. CIA Factbook: Southern ocean.
8. CIA Factbook: Arctic ocean.
9. Elert, Glenn Volume of Earth's Oceans. The Physics Factbook. Retrieved 19 April 2008.
10. Rona, Peter A. (2003). Resources of the Sea Floor. Science. 299 (5607): 673–674. doi:10.1126/science.1080679. PMID 12560541. Retrieved 2007-02-04.
11. Ralph F. Keeling, Arne Kortzinger, Nicolas Gruber (2010). Ocean Deoxygenation in a Warming World (PDF). Annual Review of Marine Science. doi:10.1146/annurev.marine.010908.163855.
12. Ocean Habitats Marietta College. Retrieved 17 April 2011.
13. Sandwell, D. T.; Smith, W. H. F. (2006-07-07). Exploring the Ocean Basins

with Satellite Altimeter Data. NOAA/NGDC. Retrieved 2007-04-21.

14. Charette, Matthew A.; Smith, Walter H. F. (June 2010). The Volume of Earth›s Ocean (PDF). Oceanography. 23 (2): 112–114. doi:10.5670/oceanog.2010.51. Retrieved 2010-06-04.
15. Ricklefs, Robert E.; Miller, Gary Leon (2000). Ecology (4th ed.). Macmillan. p. 192. ISBN 978-0-7167-2829-0.
16. Spalding, Mark, Corinna Ravilious, and Edmund Green. 2001. World Atlas of Coral Reefs. Berkeley, CA: University of California Press and UNEP/WCMC.
17. Park, Chris C. (2001). The environment: principles and applications (2nd ed.). Routledge. p. 564. ISBN 978-0-415-21770-5.
18. Davidson (2002), p.421.
19. Easterbrook (1999).
20. Garrison T (2007) Oceanography: an invitation to marine science Cengage Learning, Page 343. ISBN 978-0-495-11286-0
21. Shepard FP (1937) Revised "Classification of Marine Shorelines" The Journal of Geology, 45(6): 602–624.
22. Habitats: Beaches - Coasts Office of Naval Research. Retrieved 17 April 2011.
23. Continental shelf areas Earth trends. Retrieved 25 February 2010.
24. World The World Factbook, CIA. Retrieved 26 February 2010.
25. Moyle and Cech, 2004, page 572.
26. Hatcher, B.G. Johannes, R.E., and Robertson, A.J. (1989). Conservation of Shallow-water Marine Ecosystems. Oceanography and Marine Biology: An Annual Review. 27. Routledge. p. 320. ISBN 0-08-037718-1. Retrieved 2008-11-21.
27. Habitats: Beaches - Characteristics Office of Naval Research. Retrieved 17 April 2011.
28. Habitats: Beaches - Animal & Plant Life Office of Naval Research. Retrieved 17 April 2011.
29. Wentworth CK (1922) "A scale of grade and class terms for clastic sediments" J. Geology, 30: 377–392.
30. Tour of Rocky Shoreline Habitats Marietta College. Retrieved 17 April 2011.
31. Mangal (Mangrove). World Vegetation. Mildred E. Mathias Botanical Garden, University of California at Los Angeles
32. Morphological and Physiological Adaptations: Florida mangrove website
33. Hogarth, Peter J. (1999) The Biology of Mangroves Oxford University Press, Oxford, England, ISBN 0-19-850222-2
34. Pritchard, D. W. (1967) What is an estuary: physical viewpoint. p. 3–5 in: G. H. Lauf (ed.) Estuaries, A.A.A.S. Publ. No. 83, Washington, D.C.
35. McLusky, D.S. and Elliott, M. (2004) The Estuarine Ecosystem: ecology, threats and management. New York: Oxford University Press Inc. ISBN 0-19-852508-7
36. Wolanski, E. (2007) Estuarine Ecohydrology. Amsterdam, The Netherlands:

Elsevier. ISBN 978-0-444-53066-0

37. Bronwyn M. Gillanders, Evidence of connectivity between juvenile and adult habitats for mobile marine fauna: an important component of nurseries. 2003. Marine Ecology Progress Series
38. Jennifer A. Gill, The buffer effect and large-scale population regulation in migratory birds. 2001. Nature 412, 436-438
39. Mann, K.H. 1973. Seaweeds: their productivity and strategy for growth. Science 182: 975-981.
40. Jackson, G.A. and C.D. Winant. 1983. Effect of a kelp forest on coastal currents. Continental Shelf Report 2: 75-80.
41. Steneck, R.S., M.H. Graham, B.J. Bourque, D. Corbett, J.M. Erlandson, J.A. Estes and M.J. Tegner. 2002. Kelp forest ecosystems: biodiversity, stability, resilience and future. Environmental Conservation 29: 436-459.
42. Sala, E., C.F. Bourdouresque and M. Harmelin-Vivien. 1998. Fishing, trophic cascades, and the structure of algal assemblages: evaluation of an old but untested paradigm. Oikos 82: 425-439.
43. Dayton, P.K. 1985a. Ecology of kelp communities. Annual Review of Ecology and Systematics 16: 215-245.
44. Jones, C.G., J. H. Lawton and M. Shachak. 1997. Positive and negative effects of organisms as physical ecosystem engineers. Ecology 78: 1946-1957.
45. Druehl, L.D. 1981. The distribution of Laminariales in the North Pacific with reference to environmental influences. Proceedings of the International Congress on Systematic Evolution and Biology 2: 248-256.
46. Wheeler, W.N. 1980. Effect of boundary layer transport on the fixation of carbon by the giant kelp Macrocystis pyrifera. Marine Biology 56: 103-110.
47. Steneck, R.S. and M.N. Dethier. 1994. A functional group approach to the structure of algal-dominated communities. Oikos 69: 476-498.
48. Laffoley, Dan (December 26, 2009). To Save the Planet, Save the Seas. The New York Times. Retrieved December 2009.
49. NOAA (1998) Record-breaking coral bleaching occurred in tropics this year. National Oceanic and Atmospheric Administration, Press release (October 23, 1998).
50. ICRS (1998) Statement on Global Coral Bleaching in 1997-1998. International Coral Reef Society, October 15, 1998.
51. Bryant, D., Burke, L., McManus, J., et al. (1998) "Reefs at risk: a map-based indicator of threats to the world's coral reefs". World Resources Institute, Washington, D.C.
52. Goreau, T. J. (1992) Bleaching and Reef Community Change in Jamaica: 1951 - 1991. Am. Zool. 32: 683-695.
53. Sebens, K. P. (1994) Biodiversity of Coral Reefs: What are We Losing and Why? Am. Zool., 34: 115-133

54. Wilkinson, C. R., and Buddemeier, R. W. (1994) Global Climate Change and Coral Reefs:Implications for People and Reefs. Report of the UNEP-IOC-ASPEI-IUCN Global Task Team on the Implications of Climate Change on Coral Reefs. IUCN, Gland, Switzerland.
55. Blue Planet: Open ocean WWF›. Retrieved 17 May 2011.
56. Ocean ecology: sunlit surface waters WWF›. Retrieved 17 May 2011.
57. Moyle and Cech, 2004, page 585
58. Moyle and Cech, 2004, p. 591
59. Haedrich RL (1996) Deep-water fishes: evolution and adaptation in the earth›s largest living spaces Journal of Fish Biology 49(sA):40-53.
60. Moyle and Cech, 2004, page 586
61. McCosker, John E. (1998). Paxton, J.R.W.N., ed. Encyclopedia of Fishes. San Diego: Academic Press. p. 90. ISBN 0-12-547665-5.
62. Seven Miles Down: The Story of The Bathyscaph Trieste., Rolex Deep Sea Special, January 2006.

CHAPTER 3

Marine Microorganisms

Chapter Contents

- **Introduction**
- **Marine Viruses**
- **Prokaryotes**
- **Eukaryotes**

The term "marine microbe" covers a diversity of microorganisms, including microalgae, bacteria and archaea, protozoa fungi, and viruses. These organisms are exceedingly small—only 1/8000th the volume of a human cell and spanning about 1/100th the diameter of a human hair. Up to a million of them live in just one milliliter of seawater.

Despite being found everywhere on Earth—in your belly, dirt, the oceans—the diversity and distribution of microorganisms remain under-sampled and uncharted, especially in the deep sea.

INTRODUCTION

A microorganism (or microbe) is a microscopic living organism, which may be single-celled or multicellular. Microorganisms are very diverse and include all bacteria, archaea and most protozoa. This group also contains some species of fungi, algae, and certain microscopic animals, such as rotifers.

Many macroscopic animals and plants have microscopic juvenile stages. Some microbiologists also classify viruses (and viroids) as microorganisms, but others consider these as nonliving. In July 2016, scientists reported identifying a set of 355 genes from the last universal common ancestor (LUCA) of all life, including microorganisms, living on Earth.

Microorganisms are crucial to nutrient recycling in ecosystems as they act as decomposers. A small proportion of microorganisms are pathogenic, causing disease and even death in plants and animals. As inhabitants of the largest environment on Earth, microbial marine systems drive changes in every global system. Microbes are responsible for virtually all the photosynthesis that occurs in the ocean, as well as the cycling of carbon, nitrogen, phosphorus and other nutrients and trace elements.

Microscopic life in the oceans is very diverse and still poorly understood. For example, the role of viruses in marine ecosystems has barely being explored even in the beginning of the 21st century.

Overview

Microbial mats

Microbial mats are the earliest form of life on Earth for which there is good fossil evidence. The image shows a cyanobacterial-algal mat.

Stromatolites are formed from microbial mats as microbes slowly move upwards to avoid being smothered by sediment.

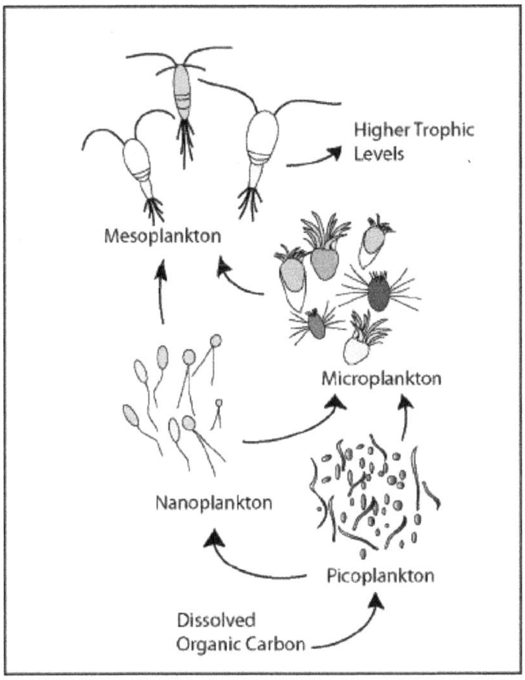

Marine microbial loop

A teaspoon of seawater contains about one million viruses. Most of these are bacteriophages, which are harmless to plants and animals, and are in fact essential to the regulation of saltwater and freshwater ecosystems. They infect and destroy bacteria in aquatic microbial communities, and are the most important mechanism of recycling carbon in the marine environment. The organic molecules released from the dead bacterial cells stimulate fresh bacterial and algal growth. Viral activity may also contribute to the biological pump, the process whereby carbon is sequestered in the deep ocean.

Marine bacteriophages are viruses that live as obligate parasitic agents in marinebacteria such as cyanobacteria. Their existence was discovered through electron microscopy and epifluorescence microscopy of ecological water samples, and later through metagenomic sampling of uncultured viral samples. The tailed bacteriophages appear to dominate marine ecosystems in number and diversity of organisms. However, viruses belonging to families Corticoviridae, Inoviridaeand Microviridae are also known to infect diverse marine bacteria. Metagenomic evidence suggests that microviruses (icosahedral ssDNA phages) are particularly prevalent in marine habitats.

Bacteriophages, viruses that are parasitic on bacteria, were first discovered in the early twentieth century. Scientists today consider that their importance in ecosystems, particularly marine ecosystems, has been underestimated, leading to these infectious agents being poorly investigated and their numbers and species biodiversity being greatly under reported.

Microorganisms constitute more than 90% of the biomass in the sea. It is estimated that viruses kill approximately 20% of this biomass each day and that there are 15 times as many viruses in the oceans as there are bacteria and archaea. Viruses are the main agents responsible for the rapid destruction of harmful algal blooms, which often kill other marine life. The number of viruses in the oceans decreases further offshore and deeper into the water, where there are fewer host organisms.

Microscopic organisms live in every part of the biosphere. The mass of prokaryote microorganisms — which includes bacteria and archaea, but not the nucleated eukaryote microorganisms — may be as much as 0.8 trillion tons of carbon (of the total biosphere mass, estimated at between 1 and 4 trillion tons). Barophilic marine microbes have been found at more than a depth of 10,000 m (33,000 ft; 6.2 mi) in the Mariana Trench, the deepest spot in the Earth›s oceans. In fact, single-celled life forms have been found in the deepest part of the Mariana Trench, by the Challenger Deep, at depths of 11,034 m (36,201 ft; 6.856 mi). Other researchers reported related studies that microorganisms thrive inside rocks up to 580 m (1,900 ft; 0.36 mi) below the sea floor under 2,590 m (8,500 ft; 1.61 mi) of ocean off the coast of the northwestern United States,as well as 2,400 m (7,900 ft; 1.5 mi) beneath the seabed off Japan. The greatest known temperature at which microbial life can exist is 122 °C (252 °F) (*Methanopyrus kandleri*). On 20 August 2014, scientists confirmed the existence of microorganisms living 800 m (2,600 ft; 0.50 mi) below the ice of Antarctica. According to one researcher, "You can find microbes everywhere — they're extremely adaptable to conditions, and survive wherever they are."

MARINE VIRUSES

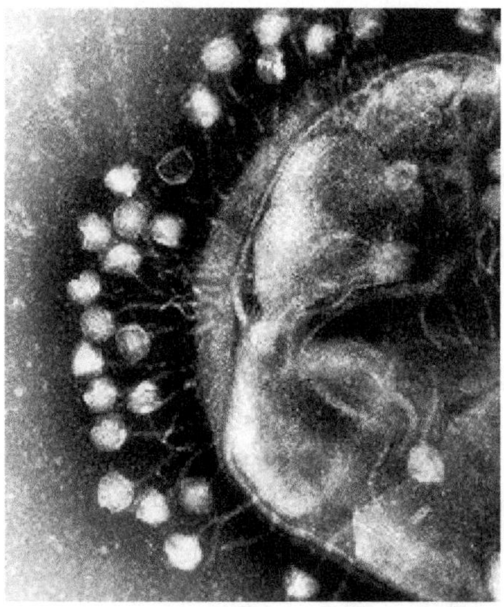

Transmission electron micrograph of multiple bacteriophages attached to a bacterial cell wall

A virus is a small infectious agent that replicates only inside the living cells of other organisms. Viruses can infect all types of life forms, from animals and plants to microorganisms, including bacteria and archaea.

When not inside an infected cell or in the process of infecting a cell, viruses exist in the form of independent particles. These viral particles, also known as *virions*, consist of two or three parts: (i) the genetic material made from either DNA or RNA, long molecules that carry genetic information; (ii) a protein coat, called the capsid, which surrounds and protects the genetic material; and in some cases (iii) an envelope of lipids that surrounds the protein coat when they are outside a cell. The shapes of these virus particles range from simple helical and icosahedral forms for some virus species to more complex structures for others. Most virus species have virions that are too small to be seen with an optical microscope. The average virion is about one one-hundredth the size of the average bacterium.

The origins of viruses in the evolutionary history of life are unclear: some may have evolved from plasmids—pieces of DNA that can move between cells—while others may have evolved from bacteria. In evolution, viruses are an important means of horizontal gene transfer, which increases genetic diversity. Viruses are considered by some to be a life form, because they carry genetic material, reproduce, and evolve through natural selection. However they lack key characteristics (such as cell structure) that are generally considered necessary to count as life. Because they possess some but not all such qualities, viruses have been described as «organisms at the edge of life» and as replicators.

Viruses are found wherever there is life and have probably existed since living cells first evolved. The origin of viruses is unclear because they do not form fossils, so molecular techniques have been used to compare the DNA or RNA of viruses and are a useful means of investigating how they arose.

Viruses are now recognised as ancient and as having origins that pre-date the divergence of life into the three domains.

Opinions differ on whether viruses are a form of life, or organic structures that interact with living organisms. They have been described as «organisms at the edge of life», since they resemble organisms in that they possess genes, evolve by natural selection, and reproduce by creating multiple copies of themselves through self-assembly. Although they have genes, they do not have a cellular structure, which is often seen as the basic unit of life. Viruses do not have their own metabolism, and require a host cell to make new products. They therefore cannot naturally reproduce outside a host cell.

Bacterial viruses, called bacteriophages, are a common and diverse group of viruses and are the most abundant form of biological entity in aquatic environments – there are up to ten times more of these viruses in the oceans than there are bacteria, reaching levels of 250,000,000 bacteriophages per millilitre of seawater.

There are also archaean viruses, which replicate within archaea: these are double-stranded DNA viruses with unusual and sometimes unique shapes. These viruses have been studied in most detail in the thermophilic archaea, particularly the orders Sulfolobales and Thermoproteales.

A teaspoon of seawater contains about one million viruses. Most of these are bacteriophages, which are harmless to plants and animals, and are in fact essential to the regulation of saltwater and freshwater ecosystems. They infect and destroy bacteria in aquatic microbial communities, and are the most important mechanism of recycling carbon in the marine environment. The organic molecules released from the dead bacterial cells stimulate fresh bacterial and algal growth. Viral activity may also contribute to the biological pump, the process whereby carbon is sequestered in the deep ocean.

Microorganisms constitute more than 90% of the biomass in the sea. It is estimated that viruses kill approximately 20% of this biomass each day and that there are 15 times as many viruses in the oceans as there are bacteria and archaea. Viruses are the main agents responsible for the rapid destruction of harmful algal blooms, which often kill other marine life. The number of viruses in the oceans decreases further offshore and deeper into the water, where there are fewer host organisms.

Viruses are an important natural means of transferring genes between different species, which increases genetic diversity and drives evolution. It is thought that viruses played a central role in the early evolution, before the diversification of bacteria, archaea and eukaryotes, at the time of the last universal common ancestor of life on Earth. Viruses are still one of the largest reservoirs of unexplored genetic diversity on Earth.

PROKARYOTES

Marine Bacteria

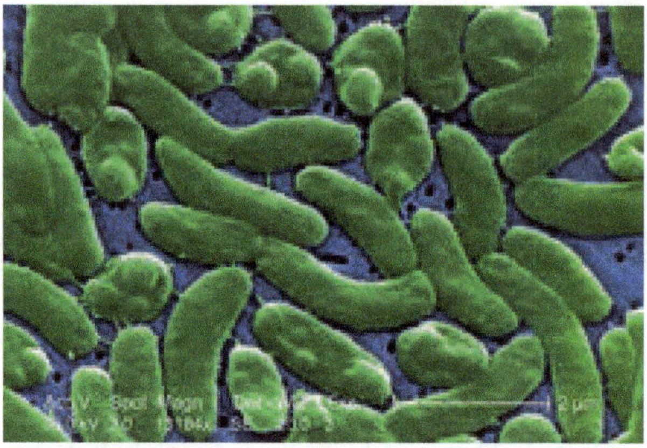

Vibrio vulnificus, a virulent bacterium found in estuaries and along coastal areas

Bacteria constitute a large domain of prokaryotic microorganisms. Typically a few micrometres in length, bacteria have a number of shapes, ranging from spheres to rods and spirals. Bacteria were among the first life forms to appear on Earth, and are present in most of its habitats. Bacteria inhabit soil, water, acidic hot springs, radioactive waste, and the deep portions of Earth›s crust. Bacteria also live in symbiotic and parasitic relationships with plants and animals.

Once regarded as plants constituting the class *Schizomycetes*, bacteria are now classified as prokaryotes. Unlike cells of animals and other eukaryotes, bacterial cells do not contain a nucleus and rarely harbour membrane-bound organelles. Although the term *bacteria* traditionally included all prokaryotes, the scientific classification changed after the discovery in the 1990s that prokaryotes consist of two very different groups of organisms that evolved from an ancient common ancestor. These evolutionary domains are called *Bacteria* and *Archaea.*

The ancestors of modern bacteria were unicellular microorganisms that were the first forms of life to appear on Earth, about 4 billion years ago. For about 3 billion years, most organisms were microscopic, and bacteria and archaea were the dominant forms of life. Although bacterial fossils exist, such as stromatolites, their lack of distinctive morphology prevents them from being used to examine the history of bacterial evolution, or to date the time of origin of a particular bacterial species. However, gene sequences can be used to reconstruct the bacterial phylogeny, and these studies indicate that bacteria diverged first from the archaeal/eukaryotic lineage. Bacteria were also involved in the second great evolutionary divergence, that of the archaea and eukaryotes. Here, eukaryotes resulted from the entering of ancient bacteria into endosymbiotic associations with the ancestors of eukaryotic cells, which were themselves possibly related to the Archaea. This involved the engulfment by proto-eukaryotic cells of alphaproteobacterial symbionts to form either mitochondria or hydrogenosomes, which are still found in all known Eukarya. Later on, some eukaryotes that already contained mitochondria also engulfed cyanobacterial-like organisms. This led to the formation of chloroplasts in algae and plants. There are also some algae that originated from even later endosymbiotic events. Here, eukaryotes engulfed a eukaryotic algae that developed into a "second-generation" plastid. This is known as secondary endosymbiosis.

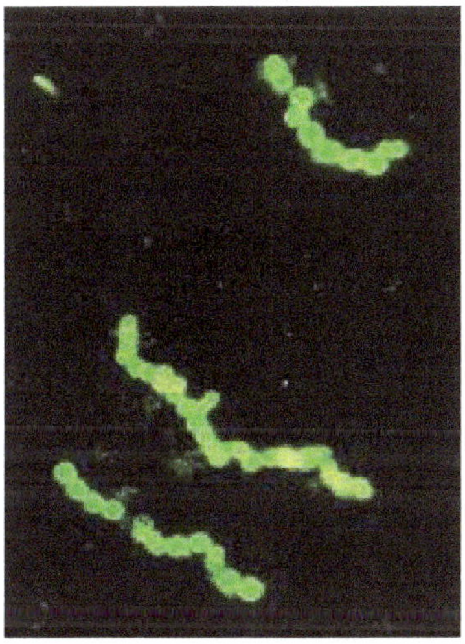

The marine *Thiomargarita namibiensis*, largest known bacterium

Cyanobacteria blooms can contain lethal cyanotoxins

Chloroplasts, such as the chloroplasts of this glaucophyte, may have originated from cyanobacteria.

Bacteria can be beneficial. This Pompeii worm, an extremophile found only at hydrothermal vents, has a protective cover of bacteria.

The largest known bacterium, the marine *Thiomargarita namibiensis*, can be visible to the naked eye and sometimes attains 0.75 mm (750 μm).

Marine Archaea

Archaea were initially viewed as extremophiles living in harsh environments, such as the yellow archaea pictured here in a hot spring, but they have since been found in a much broader range of habitats.

The archaea (Greek for *ancient*) constitute a domain and kingdom of single-celledmicroorganisms. These microbes are prokaryotes, meaning they have no cell nucleus or any other membrane-bound organelles in their cells. Archaea were initially classified as bacteria, but this classification is outdated.Archaeal cells have unique properties separating them from the other two domains of life, Bacteria and Eukaryota. The Archaea are further divided into multiple recognized phyla. Classification is difficult because the majority have not been isolated in the laboratory and have only been detected by analysis of their nucleic acids in samples from their environment.

Archaea and bacteria are generally similar in size and shape, although a few archaea have very strange shapes, such as the flat and square-shaped cells of *Haloquadratum walsbyi*. Despite this morphological similarity to bacteria, archaea possess genes and several metabolic pathways that are more closely related to those of eukaryotes, notably the enzymes involved in transcription and translation. Other aspects of archaeal biochemistry are unique, such as their reliance on ether lipids in their cell membranes, such as archaeols. Archaea use more energy sources than eukaryotes: these range from organic compounds, such as sugars, to ammonia, metal ions or even hydrogen gas. Salt-tolerant archaea (the Haloarchaea) use sunlight as an energy source, and other species of archaea fix carbon; however, unlike plants and cyanobacteria, no known species of archaea does both. Archaea reproduce asexually by binary fission, fragmentation, or budding; unlike bacteria and eukaryotes, no known species forms spores. Unlike viruses and bacteria, no known archaea is a pathogen. Archaea are particularly numerous in the oceans, and the archaea in plankton may be one of the most abundant groups of organisms on the planet. Archaea are a major part of Earth's life and may play roles in both the carbon cycle and the nitrogen cycle.

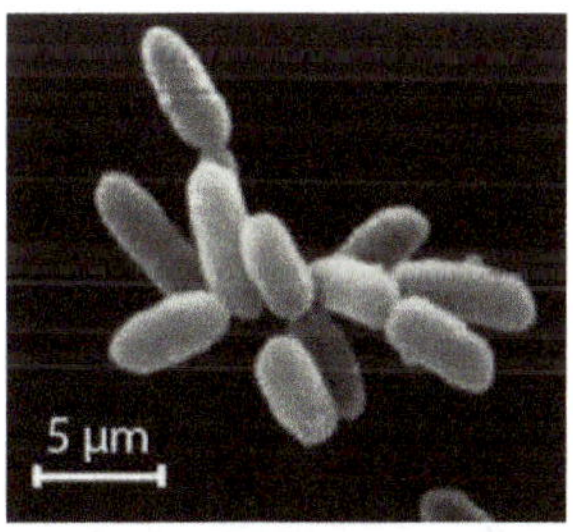

Halobacteria, found in water saturated or nearly saturated with salt, are now recognized as being archaea.

The flat and square-shaped cells of the archaea *Haloquadratum walsbyi*

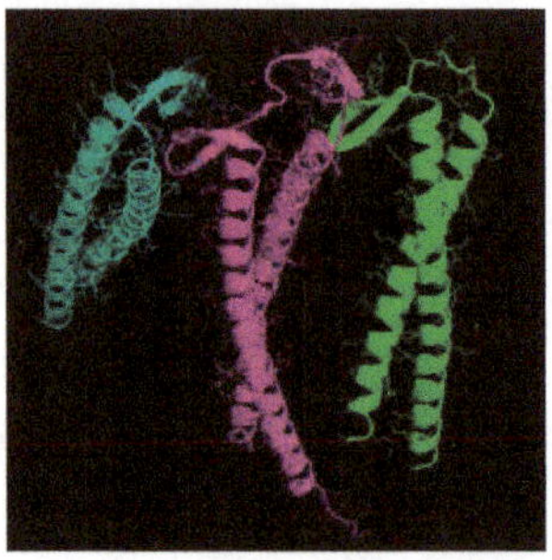

Crystal structure of prefoldin from the heat-loving marine archaea *Pyrococcus horikoshii*

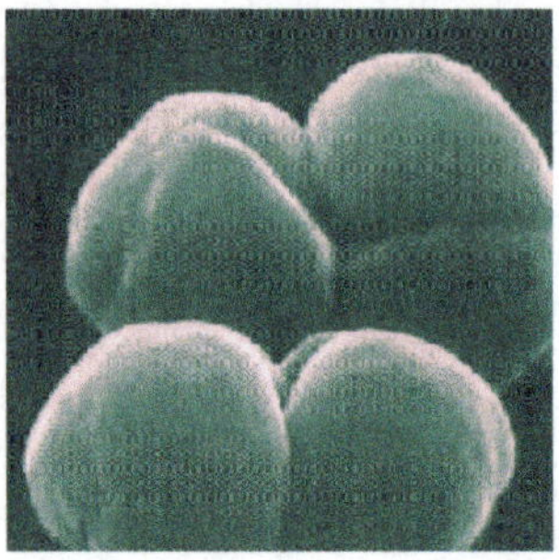

Methanosarcina barkeri, a marine archaea that produces methane

EUKARYOTES

Marine Fungi

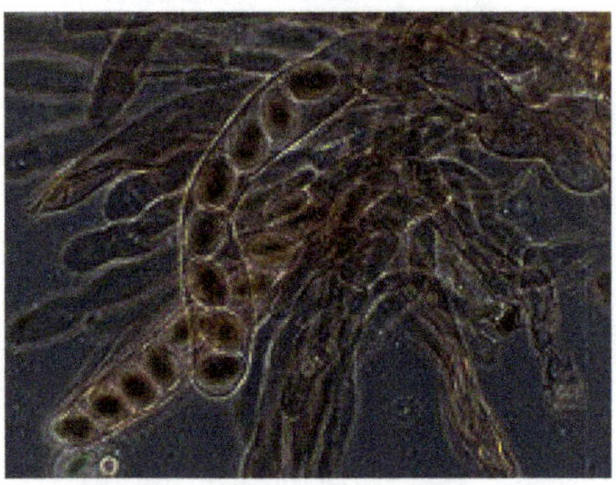

Marine Ascomycete fungus

Lichen on a rock in a marine splash zone. Lichens are mutualistic associations between a fungus and an alga or cyanobacterium.

Over 1500 species of fungi are known from marine environments. These are parasitic on marine algae or animals, or are saprobes on algae, corals, protozoan cysts, sea grasses, wood and other substrata, and can also be found in sea foam.Spores of many species have special appendages that facilitate attachment to the substratum. A very diverse range of unusual secondary metabolites is produced by marine fungi.

Mycoplankton are saprotropic members of the plankton communities of marine and freshwater ecosystems. They are composed of filamentous free-living fungi and yeasts that are associated with planktonic particles or phytoplankton. Similar to bacterioplankton, these aquatic fungi play a significant role in heterotrophicmineralization and nutrient cycling. Mycoplankton can be up to 20 mm in diameter and over 50 mm in length.

In a typical milliliter of seawater, there are approximately 10^3 to 10^4 fungal cells.This number is greater in coastal ecosystems and estuaries due to nutritional runoff from terrestrial communities. The greatest diversity and number of species of mycoplankton is found in surface waters (< 1000 m), and the vertical profile depends on the abundance of phytoplankton. Furthermore, this difference in distribution may vary between seasons due to nutrient availability. Marine fungi survive in a constant oxygen deficient environment, and therefore depend on oxygen diffusion by turbulence and oxygen generated by photosynthetic organisms.

Marine fungi can be classified as:

- Lower fungi - adapted to marine habitats (zoosporic fungi, including mastigomycetes: oomycetes and chytridiomycetes)
- Higher fungi - filamentous, modified to planktonic lifestyle (hyphomycetes, ascomycetes, basidiomycetes)

Most mycoplankton species are higher fungi.

Lichens are mutualistic associations between a fungus, usually an ascomycete, and an alga or a cyanobacterium. Several lichens are found in marine environments. Many more occur in the splash zone, where they occupy different vertical zones depending on how tolerant they are to submersion. Fossil marine lichens 600 million years old have been discovered in the late Neoproterozoic marine phosphate rocks in the sedimentary, fossil-rich Doushantuo Formation in China.

According to fossil records, fungi date back to the late Proterozoic era 900-570 million years ago. It has been hypothesized that mycoplankton evolved from terrestrial fungi, likely in the Paleozoic era (390 million years ago).

Marine Protists

Of eukaryotic groups, the protists are most commonly unicellular and microscopic. This is a highly diverse group of organisms that are not easy to classify. Several algae species are multicellular protists, and there are marine slime molds have unique life cycles that involve switching between unicellular, colonial, and multicellular forms. The number of species of protists is unknown, and we may have identified only a small portion. Studies from 2001-2004 have shown that a high degree of protist diversity exists in oceans, deep sea-vents, river sediment and an acidic river, which suggests that a large number of eukaryotic microbial communities have yet to be discovered.

Marine Algae

Algae can grow as single cells, or in long chains of cells. Green algae are a large group of photosynthetic eukaryotes that include many microscopic organisms. Green algae includes unicellular and colonial flagellates as well as various colonial, coccoid, and filamentous forms. There are about 6000 species.

- Marine algae

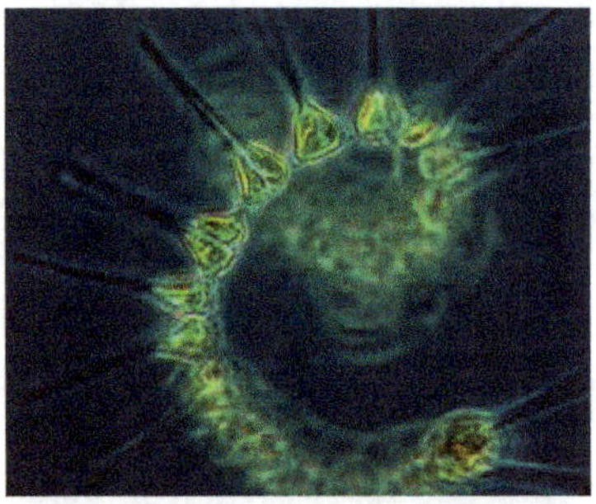

Phytoplankton are the foundation of the oceanic food chain and account for half of all photosynthetic activity on Earth.

Assorted diatoms found living between crystals of annual sea ice in Antarctica

Pleurosigma angulatum, a diatom frequently used for test purposes in microscopy

Marine Animals

At least one microscopic animal group, the parasitic cnidarian Myxozoa, is unicellular in its adult form, and includes marine species. Other marine micro-animals are multicellular. Microscopic arthropods are more commonly found inland in freshwater, but there are marine species as well. Microscopic marine crustaceans include some copepods, cladocera and water bears. Some marine nematodes and rotifers are also too small to be seen with the naked eye, as are many loricifera, including the recently discovered anaerobic species that spend their lives in an anoxic environment. Copepods contribute more to the secondary productivity and carbon sink of the world oceans than any other group of organisms.

- Marine micro-animals

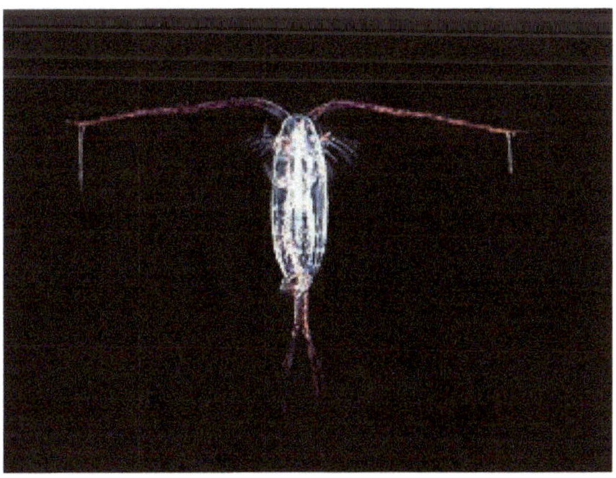

Over 10,000 marine species are copepods, small, often microscopic crustaceans

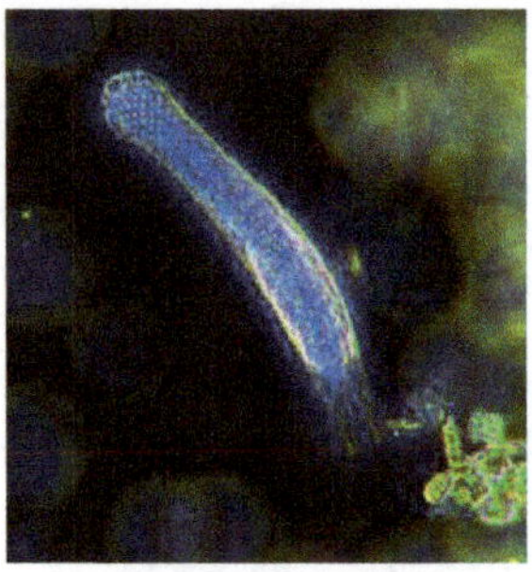

Darkfield photo of a gastrotrich, which form a phylum of worm-like animals living mostly between particles of sediment

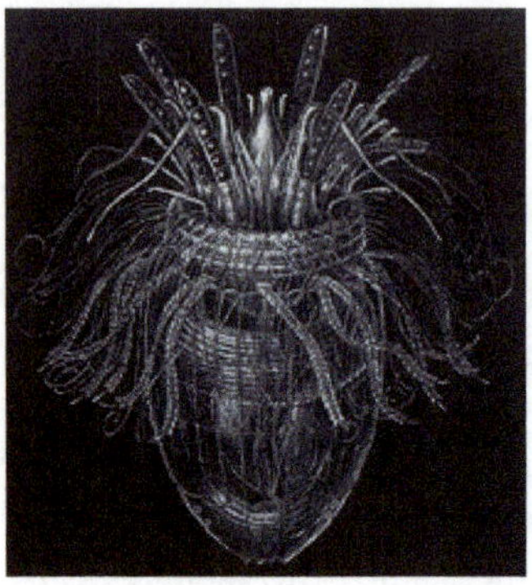

Armoured *Pliciloricus enigmaticus* of the phylum Loricifera live in the spaces between marine gravel

REFERENCES

1. Alexander, David E. (1 May 1999). Encyclopedia of Environmental Science. Springer. ISBN 0-412-74050-8.
2. United States Environmental Protection Agency (2 March 2006). «Marine Ecosystems». Retrieved 2006-08-25.
3. Daily, Gretchen C. (1 February 1997). Nature›s Services. Island Press. ISBN 1-55963-476-6.
4. Vaccari, David A. (8 November 2005). Environmental Biology for Engineers and Scientists. Wiley-Interscience. ISBN 0-471-74178-7.
5. Clegg, J. (1986). Observer's Book of Pond Life. Frederick Warne, London. 460 p.
6. Clegg, J. (1986). Observer's Book of Pond Life. Frederick Warne, London. 460 p. p.160-163.
7. Keddy, Paul A. (2010). Wetland Ecology. Principles and Conservation. Cambridge University Press. p. 497. ISBN 978-0-521-51940-3.

8. Keddy, P.A., D. Campbell, T. McFalls, G. Shaffer, R. Moreau, C. Dranguet, and R. Heleniak. 2007. The wetlands of lakes Pontchartrain and Maurepas: past, present and future. Environmental Reviews 15: 1- 35.
9. Gastescu, P. (1993). The Danube Delta: geographical characteristics and ecological recovery. Earth and Environmental Science, 29, 57–67.
10. Loeb, Stanford L. (24 January 1994). Biological Monitoring of Aquatic Systems. CRC Press. ISBN 0-87371-910-7.
11. Vallentyne, J. R. (1974). The Algal Bowl: Lakes and Man, Miscellaneous Special Publication No. 22. Ottawa, ON: Department of the Environment, Fisheries and Marine Service.
12. Moss, B. (1983). The Norfolk Broadland: experiments in the restoration of a complex wetland. Biological Reviews of the Cambridge Philosophical Society, 58, 521–561.
13. Keddy, P. A., Campbell, D., McFalls T., Shaffer, G., Moreau, R., Dranguet, C., and Heleniak, R. (2007). The wetlands of lakes Pontchartrain and Maurepas: past, present and future. Environmental Reviews, 15, 1–35.
14. Graham, J. B. (1997). Air Breathing Fishes. San Diego, CA: Academic Press.
15. Sculthorpe, C. D. (1967). The Biology of Aquatic Vascular Plants. Reprinted 1985 Edward Arnold, by London.
16. Manahan, Stanley E. (1 January 2005). Environmental Chemistry. CRC Press. ISBN 1-56670-633-5.
17. Smith, V. H. (1982). The nitrogen and phosphorus dependence of algal biomass in lakes: an empirical and theoretical analysis. Limnology and Oceanography, 27, 1101–12.
18. Smith, V. H. (1983). Low nitrogen to phosphorus ratios favor dominance by bluegreen algae in lake phytoplankton. Science, 221, 669–71.
19. Turner, R. E. and Rabelais, N. N. (2003). Linking landscape and water quality in the Mississippi River Basin for 200 years. BioScience, 53, 563–72.
20. Silliman, B. R., Grosholz, E. D., and Bertness, M. D. (eds.) (2009). Human Impacts on Salt Marshes: A Global Perspective. Berkeley, CA: University of California Press.
21. Chapman, J.L.; Reiss, M.J. (10 December 1998). Ecology. Cambridge University Press. ISBN 0-521-58802-2.

CHAPTER 4

Marine Invertebrates

Chapter Contents

- **Introduction**
- **Evolution**
- **Classification**
- **Marine Cnidarians**
- **Marine Worms**
- **Echinoderms**
- **Marine Molluscs**
- **Marine Arthropods**
- **Other Phyla**
- **Minerals From Sea Water**

Invertebrates are animals that do not have backbones. Many are only found in the sea. Marine invertebrates are a extremely large and diverse group that includes corals, sea anemones, sea urchins, starfish and many crustaceans, molluscs and worms. Crustaceans and molluscs are sometimes grouped together and called "shellfish". Among the marine invertebrates are some of the largest animals on Earth (the colossal squid may measure up to 14 metres [46 feet long]) as well as some of the deadliest

INTRODUCTION

Marine invertebrates are the invertebrates that live in marine habitats. Invertebrate is a blanket term that includes all animals apart from the vertebrate members of the chordate phylum. Invertebrates lack a vertebral column, and some have evolved a shell or a hard exoskeleton. As on land and in the air, marine invertebrates have a large variety of body plans, and have been categorised into over 30 phyla. They make up most of the macroscopic life in the oceans.

Ernst Haeckel's 96th plate, showing some marine invertebrates. Marine invertebrates have a large variety of body plans, which are currently categorised into over 30 phyla.

EVOLUTION

Kimberella, an early mollusc important for understanding the Cambrian explosion. Invertebrates are grouped into different phyla (body plans).

Opabinia an extinct, stem grouparthropod that appeared in the Middle Cambrian

The earliest animals were marine invertebrates, that is, vertebrates came later. Animals are multicellular eukaryotes, and are distinguished from plants, algae, and fungi by lacking cell walls. Marine invertebrates are animals that inhabit a marine environment apart from the vertebrate members of the chordate phylum; invertebrates lack a vertebral column. Some have evolved a shell or a hard exoskeleton.

The earliest widely accepted animal fossils are the rather modern-looking cnidarians (the group that includes jellyfish, sea anemones and *Hydra*), possibly from around 580 Ma The Ediacara biota, which flourished for the last 40 million years before the start of the Cambrian, were the first animals more than a very few centimetres long. Many were flat and had a quilted appearance, and seemed so strange that there was a proposal to classify them as a separate kingdom, Vendozoa. Others, however, have been interpreted as early molluscs (*Kimberella*), echinoderms (*Arkarua*), and arthropods (*Spriggina*, *Parvancorina*). There is still debate about the classification of these specimens, mainly because the diagnostic features which allow taxonomists to classify more recent organisms, such as similarities to living organisms, are generally absent in the Ediacarans. However, there seems little doubt that *Kimberella* was at least a triploblastic bilaterian animal, in other words, an animal significantly more complex than the cnidarians.

The small shelly fauna are a very mixed collection of fossils found between the Late Ediacaran and Middle Cambrian periods. The earliest, *Cloudina*, shows signs of successful defense against predation and may indicate the start of an evolutionary arms race. Some tiny Early Cambrian shells almost certainly belonged to molluscs, while the owners of some "armor plates," *Halkieria* and *Microdictyon*, were eventually identified when more complete specimens were found in Cambrian lagerstätten that preserved soft-bodied animals.

In the 1970s there was already a debate about whether the emergence of the modern phyla was explosive or gradual but hidden by the shortage of Precambrian animal fossils. A re-analysis of fossils from the Burgess Shale lagerstätte increased interest in the issue when it revealed animals, such as *Opabinia*, which did not fit into any known phylum. At the time these were interpreted as evidence that the modern phyla had evolved very rapidly in the Cambrian explosion and that the Burgess Shale's "weird wonders" showed that the Early Cambrian was a uniquely experimental period of animal evolution. Later discoveries of similar animals and the development of new theoretical approaches led to the conclusion that many of the "weird wonders" were evolutionary "aunts" or "cousins" of modern groups—for example that *Opabinia* was a member of the lobopods, a group which includes the ancestors of the arthropods, and that it may have been closely related to the modern tardigrades. Nevertheless, there is still much debate about whether the Cambrian explosion was really explosive and, if so, how and why it happened and why it appears unique in the history of animals.

CLASSIFICATION

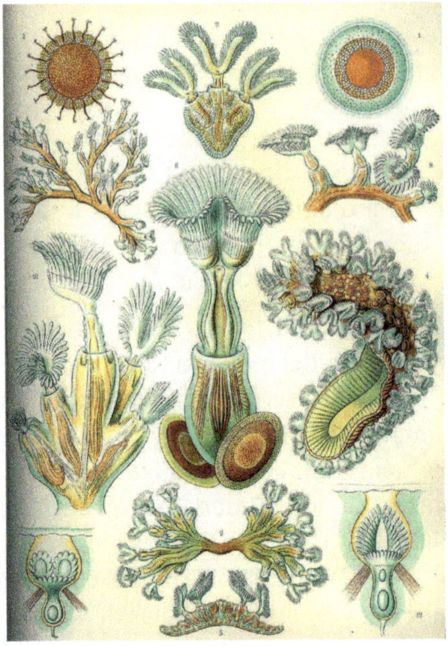

Bryozoa, from Ernst Haeckel's *Kunstformen der Natur*, 1904

Invertebrates are grouped into different phyla. Informally phyla can be thought of as a way of grouping organisms according to their body plan.[:33] A body plan refers to a blueprint which describes the shape or morphology of an organism, such as its symmetry, segmentation and the disposition of its appendages. The idea of body plans originated with vertebrates, which were grouped into one phylum. But the vertebrate body plan is only one of many, and invertebrates consist of many phyla or body plans. The history of the discovery of body plans can be seen as a movement from a worldview centred on vertebrates, to seeing the vertebrates as one body plan among many. Among the pioneering zoologists, Linnaeus identified two body plans outside the vertebrates; Cuvier identified three; and Haeckel had four, as well as the Protista with eight more, for a total of twelve. For comparison, the number of phyla recognised by modern zoologists has risen to 35.

Historically body plans were thought of as having evolved in rapidly during the Cambrian explosion, but a more nuanced understanding of animal evolution suggests a gradual development of body plans throughout the early Palaeozoic and beyond. More generally a phylum can be defined in two ways: as described above, as a group of organisms with a certain degree of morphological or developmental similarity (the phenetic definition), or a group of organisms with a certain degree of evolutionary relatedness (the phylogenetic definition).

As on land and in the air, invertebrates make up a great majority of all macroscopic life, as the vertebrates makes up a subphylum of one of over 30 known animal phyla, making the term almost meaningless for taxonomic purpose. Invertebrate sea life includes the following groups, some of which are phyla:

The 49th plate from Ernst Haeckel's *Kunstformen der Natur*, 1904, showing various sea anemones classified as Actiniae, in the Cnidaria phylum

"A variety of marine worms": plate from *Das Meer* by M.J. Schleiden (1804–1881)

- Acoela, among the most primitive bilateral animals.
- Annelida, (polychaetes and sea leeches).
- Brachiopoda, marine animals that have hard valves (shells) on the upper and lower surfaces.
- Bryozoa, also known as moss animals or sea mats.
- Chaetognatha, commonly known as arrow worms, are a phylum of predatory marine worms that are a major component of plankton.
- Cephalochordata represented in the modern oceans by the lancelets (also known as Amphioxus).
- Cnidaria, such as jellyfish, sea anemones, and corals.
- Crustacea, including lobsters, crabs, shrimp, crayfish, barnacles, hermit crabs, mantis shrimps, and copepods.
- Ctenophora, also known as comb jellies, the largest animals that swim by means of cilia.
- Echinodermata, including sea stars, brittle stars, sea urchins, sand dollars, sea cucumbers, crinoids, and sea daisies.
- Echiura, also known as spoon worms.
- Gnathostomulids, slender to thread-like worms, with a transparent body that inhabit sand and mud beneath shallow coastal waters.
- Gastrotricha, often called hairy backs, found mostly interstitially in between sediment particles.

- Hemichordata, includes acorn worms, solitary worm-shaped organisms.
- Kamptozoa, goblet-shaped sessile aquatic animals, with relatively long stalks and a crown of solid tentacles, also called Entoprocta.
- Kinorhyncha, segmented, limbless animals, widespread in mud or sand at all depths, also called mud dragons.
- Loricifera, very small to microscopic marine sediment-dwelling animals only discovered in 1983.
- Merostomata; also known as horseshoe crabs.
- Mollusca, including shellfish, squid, octopus, whelks, *Nautilus*, cuttlefish, nudibranchs, scallops, sea snails, Aplacophora, Caudofoveata, Monoplacophora, Polyplacophora, and Scaphopoda.
- Myzostomida, a taxonomic group of small marine worms which are parasitic on crinoids or "sea lilies".
- Nemertinea, also known as "ribbon worms" or "proboscis worms".
- Orthonectida, a small phylum of poorly known parasites of marine invertebrates that are among the simplest of multi-cellular organisms.
- Phoronida, a phylum of marine animals that filter-feed with a lophophore (a "crown" of tentacles), and build upright tubes of chitin to support and protect their soft bodies.
- Placozoa, small, flattened, multicellular animals around 1 millimetre across and the simplest in structure. They have no regular outline, although the lower surface is somewhat concave, and the upper surface is always flattened.
- Porifera (sponges), multicellular organisms that have bodies full of pores and channels allowing water to circulate through them.
- Priapulida, or penis worms, are a phylum of marine worms that live marine mud. They are named for their extensible spiny proboscis, which, in some species, may have a shape like that of a human penis.
- Pycnogonida, also called sea spiders, are unrelated to spiders, or even to arachnids which they resemble.
- Sipunculida, also called peanut worms, is a group containing 144–320 species (estimates vary) of bilaterally symmetrical, unsegmented marine worms.
- Tunicata, also known as sea squirts or sea pork, are filter feeders attached to rocks or similarly suitable surfaces on the ocean floor.
- Some flatworms of the classes Turbellaria and Monogenea.
- Xenoturbella, a genus of bilaterian animals that contains only two marine worm-like species.
- Xiphosura, includes a large number of extinct lineages and only four recent species in the family Limulidae, which include the horseshoe crabs.

Arthropods total about 1,113,000 described extant species, molluscs about 85,000 and chordates about 52,000.

MARINE SPONGES

Sponges have no nervous, digestive or circulatory system

Sponges are animals of the phylum Porifera (Modern Latin for *bearing pores*). They are multicellular organisms that have bodies full of pores and channels allowing water to circulate through them, consisting of jelly-like mesohyl sandwiched between two thin layers of cells. They have unspecialized cells that can transform into other types and that often migrate between the main cell layers and the mesohyl in the process. Sponges do not have nervous, digestive or circulatory systems. Instead, most rely on maintaining a constant water flow through their bodies to obtain food and oxygen and to remove wastes.

Sponges are similar to other animals in that they are multicellular, heterotrophic, lack cell walls and produce sperm cells. Unlike other animals, they lack true tissues and organs, and have no body symmetry. The shapes of their bodies are adapted for maximal efficiency of water flow through the central cavity, where it deposits nutrients, and leaves through a hole called the osculum. Many sponges have internal skeletons of spongin and/or spicules of calcium carbonate or silicon dioxide. All sponges are sessile aquatic animals. Although there are freshwater species, the great majority are marine (salt water) species, ranging from tidal zones to depths exceeding 8,800 m (5.5 mi).

While most of the approximately 5,000–10,000 known species feed on bacteria and other food particles in the water, some host photosynthesizing micro-organisms as endosymbionts and these alliances often produce more food and oxygen than they consume. A few species of sponge that live in food-poor environments have become carnivores that prey mainly on small crustaceans.

Sponge biodiversity. There are four sponge species in this photo.

Branching vase sponge

Venus' flower basket at a depth of 2572 meters

Barrel sponge

Stove-pipe sponge

Linnaeus mistakenly identified sponges as plants in the order Algae. For a long time thereafter sponges were assigned to a separate subkingdom, Parazoa (meaning *beside the animals*). They are now classified as a paraphyletic phylum from which the higher animals have evolved.

MARINE CNIDARIANS

Cnidarians are the simplest animals with cells organised into tissues. Yet the starlet sea anemone contains the same genes as those that form the vertebrate head.

Cnidarians (Greek for *nettle*) are distinguished by the presence of stinging cells, specialized cells that they use mainly for capturing prey. Cnidarians include corals, sea anemones, jellyfish and hydrozoans. They form a phylum containing over 10,000species of animals found exclusively in aquatic (mainly marine) environments. Their bodies consist of mesoglea, a non-living jelly-like substance, sandwiched between two layers of epithelium that are mostly one cell thick. They have two basic body forms: swimming medusae and sessile polyps, both of which are radially symmetrical with mouths surrounded by tentacles that bear cnidocytes. Both forms have a single orifice and body cavity that are used for digestion and respiration.

Fossil cnidarians have been found in rocks formed about 580 million years ago. Fossils of cnidarians that do not build mineralized structures are rare. Scientists currently think cnidarians, ctenophores and bilaterians are more closely related to calcareous sponges than these are to other sponges, and that anthozoans are the evolutionary "aunts" or "sisters" of other cnidarians, and the most closely related to bilaterians.

Cnidarians are the simplest animals in which the cells are organised into tissues. The starlet sea anemone is used as a model organism in research. It is easy to care for in the laboratory and a protocol has been developed which can yield large numbers of embryos on a daily basis. There is a remarkable degree of similarity in the gene sequence conservation and complexity between the sea anemone and vertebrates. In particular, genes concerned in the formation of the head in vertebrates are also present in the anemone.

Sea anemones are common in tidepools

Their tentacles sting and paralyse small fish

Close up of polyps on the surface of a coral, waving their tentacles.

As islands sink below the sea, corals growth can keep pace with the rising water, forming an atoll

Portuguese man o' war

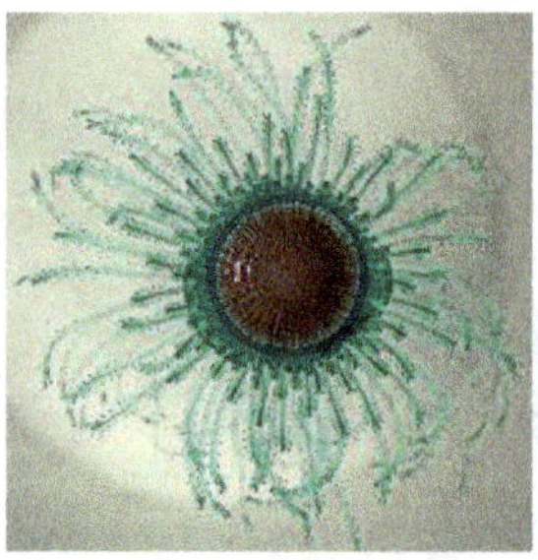

Porpita porpita

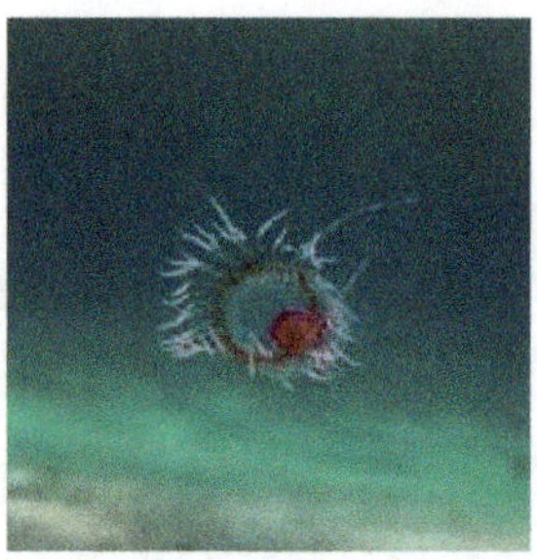

Turritopsis dohrnii, a small, biologically immortal jellyfish transfers its cells back to childhood.

Moon jellyfish, found in coastal waters around the world

Lion's mane jellyfish, largest known jellyfish

MARINE WORMS

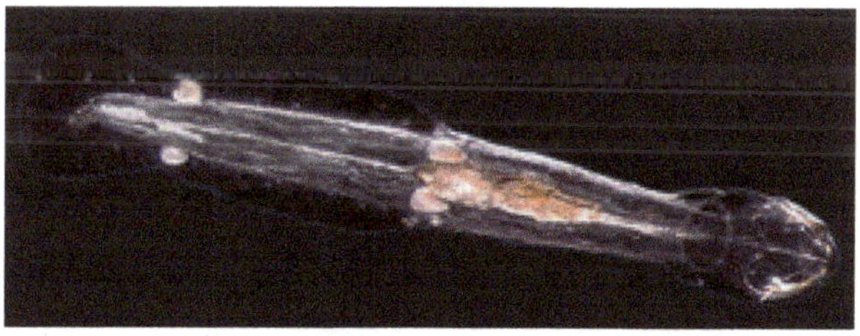

Arrow worms are predatory components of plankton worldwide.

Worms (Old English for *serpent*) typically have long cylindrical tube-like bodies and no limbs. Marine worms vary in size from microscopic to over 1 metre (3.3 ft) in length for some marine polychaete worms (bristle worms) and up to 58 metres (190 ft) for the marine nemertean worm (bootlace worm). Some marine worms occupy a small variety of parasitic niches, living inside the bodies of other animals, while others live more freely in the marine environment or by burrowing underground.

Different groups of marine worms are related only distantly, so they are found in several different phyla such as the Annelida (segmented worms), Chaetognatha (arrow worms), Hemichordata, and Phoronida (horseshoe worms). Many of these worms have specialized tentacles used for exchanging oxygen and carbon dioxide and also may be used for reproduction. Some marine worms are tube worms, such as the giant tube worm which lives in waters near underwater volcanoes and can withstand temperatures up to 90 degrees Celsius.

Platyhelminthes (flatworms) form another worm phylum which includes a class Cestoda of parasitic tapeworms. The marine tapeworm *Polygonoporus giganticus*, found in the gut of sperm whales, can grow to over 30 m (100 ft).

Nematodes (roundworms) constitute a further worm phylum with tubular digestive systems and an opening at both ends.Over 25,000 nematode species have been described, of which more than half are parasitic, t, and it has been estimated another million remain undescribed. They are ubiquitous in marine, freshwater and terrestrial environments, where they often outnumber other animals in both individual and species counts. They are found in every part of the earth's lithosphere, from the top of mountains to the bottom of oceanic trenches. By count they represent 90% of all animals on the ocean floor. Their numerical dominance, often exceeding a million individuals per square meter and accounting for about 80% of all individual animals on earth, their diversity of life cycles, and their presence at various trophic levels point at an important role in many ecosystems.

The bootlace worm can grow to 58 metres (190 ft).

Giant tube worms cluster around hydrothermal vents

Lamellibrachia luymes, a cold seep tubeworm, lives over 250 years.

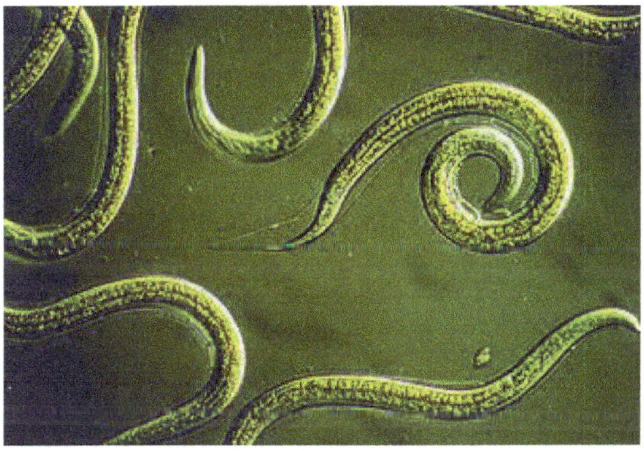

Nematodes are ubiquitous pseudocoelomates which can parasite marine plants and animals.

Bloodworms are typically found on the bottom of shallow marine waters

ECHINODERMS

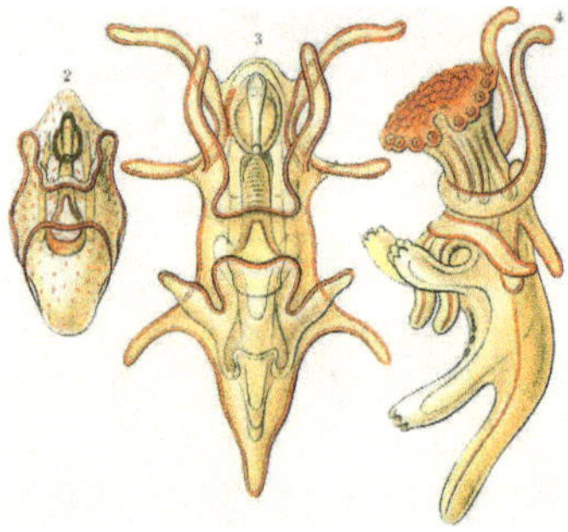

Starfish larvae are bilaterally symmetric, whereas the adults have fivefold symmetry

Echinoderms (Greek for *spiny skin*) is a phylum which contains only marine invertebrates. The adults are recognizable by their radial symmetry (usually five-point) and include starfish, sea urchins, sand dollars, and sea cucumbers, as well as the sea lilies. Echinoderms are found at every ocean depth, from the intertidal zone to the abyssal zone. The phylum contains about 7000 living species, making it the second-largest grouping of deuterostomes (a superphylum), after the chordates (which include the vertebrates, such as birds, fishes, mammals, and reptiles).

Echinoderms are unique among animals in having bilateral symmetry at the larval stage, but fivefold symmetry (pentamerism, a special type of radial symmetry) as adults.

The echinoderms are important both biologically and geologically. Biologically, there are few other groupings so abundant in the biotic desert of the deep sea, as well as shallower oceans. The more notably distinct trait, which most echinoderms have, is their remarkable powers of regeneration of tissue, organs, limbs, and of asexual reproduction, and in some cases, complete regeneration from a single limb. Geologically, the value of echinoderms is in their ossified skeletons, which are major contributors to many limestone formations, and can provide valuable clues as to the geological environment. They were the most used species in regenerative research in the 19th and 20th centuries. Further, it is held by some scientists that the radiation of echinoderms was responsible for the Mesozoic Marine Revolution.

Echinoderm literally means "spiny skin", as this water melon sea urchin illustrates

Sea cucumbers filter feed on plankton and suspended solids

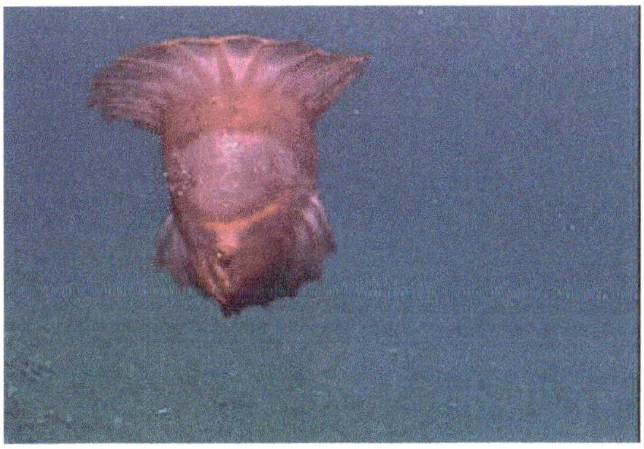

Benthopelagic sea cucumbers can lift off the seafloor and journey as much as 1,000 m (3,300 ft) up the water column

The ochre sea star was the first keystone predator to be studied. They limit mussels which can overwhelm intertidal communities.

Colorful sea lilies in shallow waters

Aside from the hard-to-classify *Arkarua* (a Precambrian animal with echinoderm-like pentamerous radial symmetry), the first definitive members of the phylum appeared near the start of the Cambrian.

MARINE MOLLUSCS

Reconstruction of an ammonite, a highly successful early cephalopod that first appeared in the Devonian (about 400 mya). They became extinct during the same extinction event that killed the land dinosaurs (about 66 mya).

Molluscs (Latin for *soft*) form a phylum with about 85,000 extant recognized species. They are the largest marine phylum, comprising about 23% of all the named marine organisms. Molluscs have more varied forms than other invertebrate phylums. They are highly diverse, not just in size and in anatomical structure, but also in behaviour and in habitat. The majority of species still live in the oceans, from the seashores to the abyssal zone, but some form a significant part of the freshwater fauna and the terrestrial ecosystems.

The mollusc phylum is divided into 9 or 10 taxonomic classes, two of which are extinct. These classes include gastropods, bivalves and cephalopods, as well as other lesser-known but distinctive classes. Gastropods with protective shells are referred to as snails, whereas gastropods without protective shells are referred to as slugs. Gastropods are by far the most numerous molluscs in terms of classified species, accounting for 80% of the total. Bivalves include clams, oysters, cockles, mussels, scallops, and numerous other families. There are about 8,000 marine bivalves species (including brackish water and estuarine species), and about 1,200 freshwater species. Cephalopod include octopus, squid and cuttlefish. They are found in all oceans, and neurologically are the most advanced of the invertebrates. About 800 living species of marine cephalopods have been identified, and an estimated 11,000 extinct taxa have been described. There are no fully freshwater cephalopods.

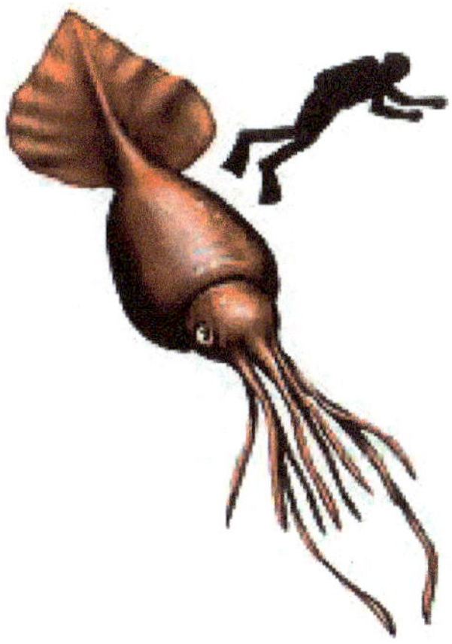

Colossal squid, largest of all invertebrates

The nautilus is a living fossil little changed since it evolved 500 million years ago as one of the first cephalopods.

Marine gastropods are sea snails or sea slugs. This nudibranch is a sea slug.

The sea snail *Syrinx aruanus* has the largest shell of any living gastropod

Molluscs usually have eyes. Bordering the edge of the mantle of a scallop, a bivalve mollusc, can be over 100 simple eyes.

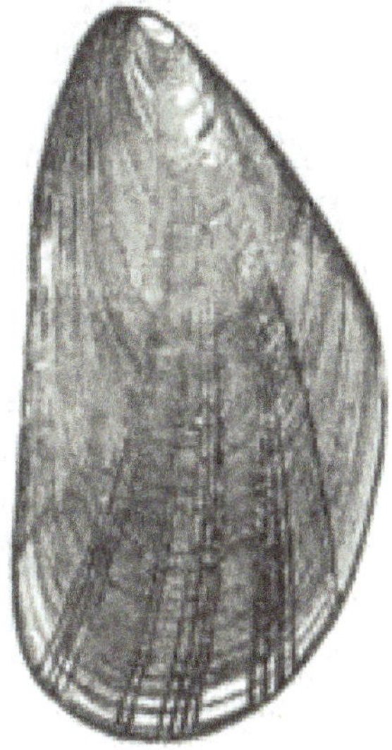

Common mussel, another bivalve

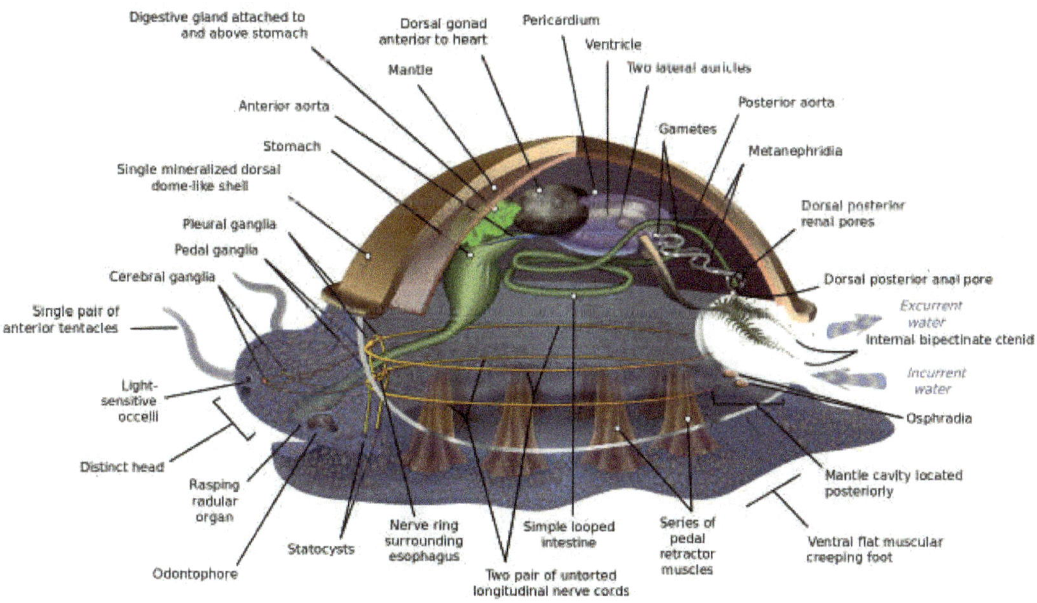

Generalized or *hypothetical ancestral mollusc*

Molluscs have such diverse shapes that many textbooks base their descriptions of molluscan anatomy on a generalized or *hypothetical ancestral mollusc*. This generalized mollusc is unsegmented and bilaterally symmetrical with an underside consisting of a single muscular foot. Beyond that it has three further key features. Firstly, it has a muscular cloak called a mantle covering its viscera and containing a significant cavity used for breathing and excretion. A shell secreted by the mantle covers the upper surface. Secondly (apart from bivalves) it has a rasping tongue called a radula used for feeding. Thirdly, it has a nervous system including a complex digestive system using microscopic, muscle-powered hairs called cilia to exude mucus. The generalized mollusc has two paired nerve cords (three in bivalves). The brain, in species that have one, encircles the esophagus. Most molluscs have eyes and all have sensors detecting chemicals, vibrations, and touch. The simplest type of molluscan reproductive system relies on external fertilization, but more complex variations occur. All produce eggs, from which may emerge trochophore larvae, more complex veliger larvae, or miniature adults. The depiction is rather similar to modern monoplacophorans, and some suggest it may resemble very early molluscs.

Good evidence exists for the appearance of marine gastropods, cephalopods and bivalves in the Cambrian period 541 to 485.4 million years ago. However, the evolutionary history both of molluscs› emergence from the ancestral Lophotrochozoa and of their diversification into the well-known living and fossil forms are still subjects of vigorous debate among scientists.

MARINE ARTHROPODS

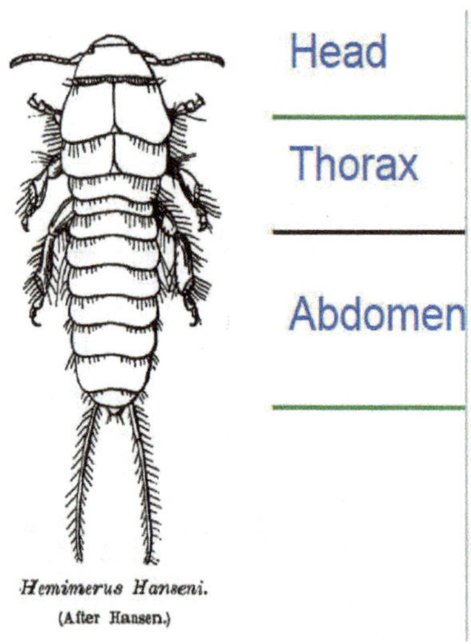

Segmentation and tagmata of an arthropod

Arthropods (Greek for *jointed feet*) have an exoskeleton (external skeleton), a segmented body, and jointed appendages (paired appendages). They form a phylum which includes insects, arachnids, myriapods, and crustaceans. Arthropods are characterized by their jointed limbs and cuticle made of chitin, often mineralised with calcium carbonate. The arthropod body plan consists of segments, each with a pair of appendages. The rigid cuticle inhibits growth, so arthropods replace it periodically by moulting. Their versatility has enabled them to become the most species-rich members of all ecological guilds in most environments.

Marine arthropods range in size from the microscopic crustacean *Stygotantulus* to the Japanese spider crab. Arthropods› primary internal cavity is a hemocoel, which accommodates their internal organs, and through which their haemolymph - analogue of blood - circulates; they have open circulatory systems. Like their exteriors, the internal organs of arthropods are generally built of repeated segments. Their nervous system is "ladder-like", with paired ventral nerve cords running through all segments and forming paired ganglia in each segment. Their heads are formed by fusion of varying numbers of segments, and their brains are formed by fusion of the ganglia of these segments and encircle the esophagus. The respiratory and excretory systems of arthropods vary, depending as much on their environment as on the subphylum to which they belong.

Their vision relies on various combinations of compound eyes and pigment-pit ocelli: in most species the ocelli can only detect the direction from which light is coming, and the compound eyes are the main source of information, but the main eyes of spiders are ocelli that can form images and, in a few cases, can swivel to track prey. Arthropods also have a wide range of chemical and mechanical sensors, mostly based on modifications of the many setae (bristles) that project through their cuticles. Arthropods' methods of reproduction and development are diverse; all terrestrial species use internal fertilization, but this is often by indirect transfer of the sperm via an appendage or the ground, rather than by direct injection. Marine species all lay eggs and use either internal or external fertilization. Arthropod hatchlings vary from miniature adults to grubs that lack jointed limbs and eventually undergo a total metamorphosis to produce the adult form.

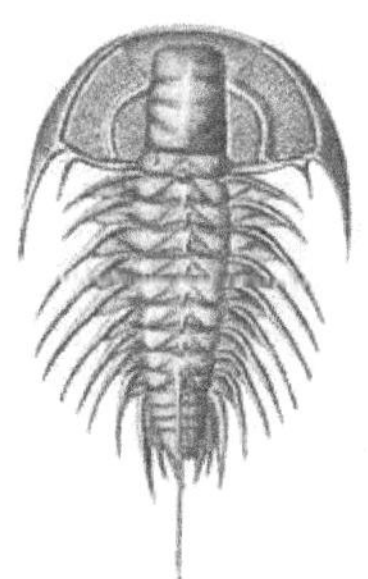

Trilobites, now extinct, roamed oceans for 270 million years.

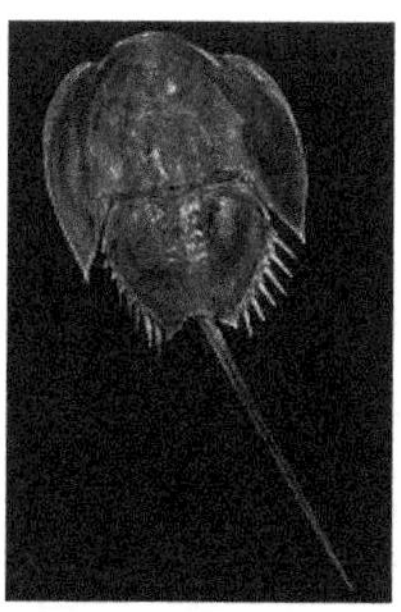

Horseshoe crab, a living fossil arthropod from 450 million years ago

- Crustaceans

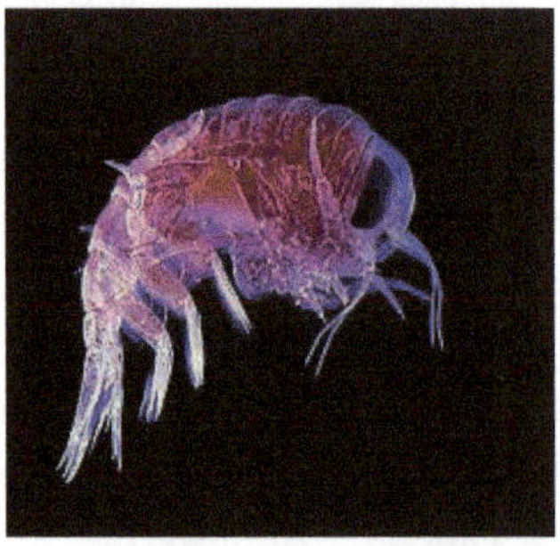

Many crustaceans are very small, like this tiny amphipod, and make up a significant part of the ocean's zooplankton

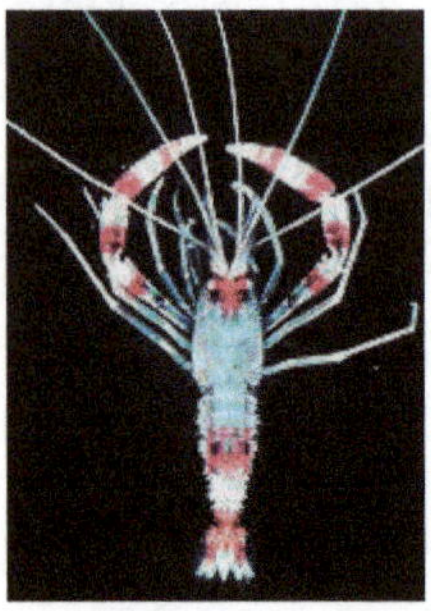

The banded cleaner shrimp is a crustacean common in the tropics.

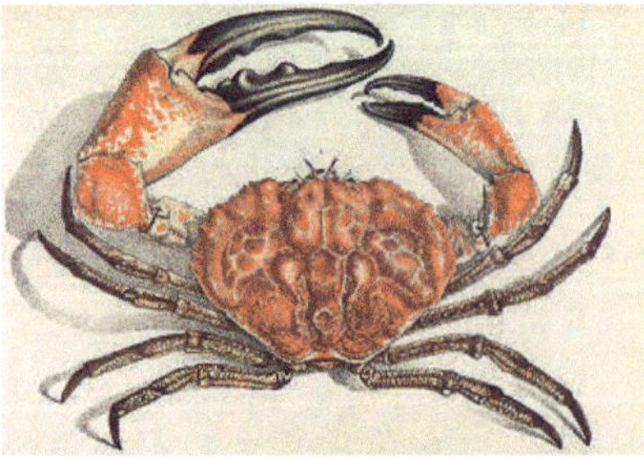

The Tasmanian giant crab is long-lived and slow-growing, making it vulnerable to overfishing.

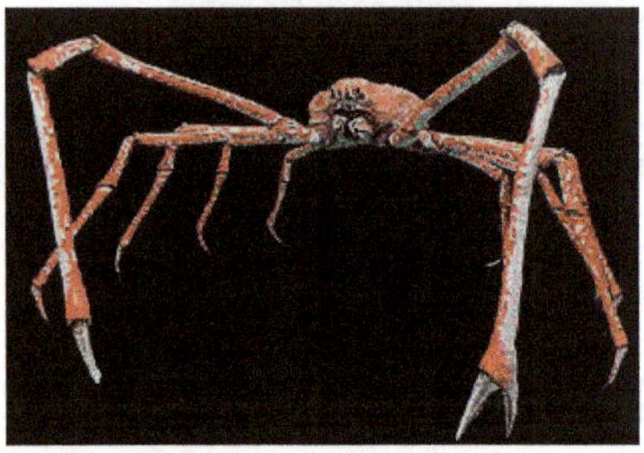

The Japanese spider crab has the longest leg span of any arthropod.

The evolutionary ancestry of arthropods dates back to the Cambrian period. The group is generally regarded as monophyletic, and many analyses support the placement of arthropods with cycloneuralians (or their constituent clades) in a superphylum Ecdysozoa. Overall however, the basal relationships of Metazoa are not yet well resolved. Likewise, the relationships between various arthropod groups are still actively debated.

OTHER PHYLA

- Tardigrade, Lobopodia, (Onychophora)
- Non-craniate (non-vertebrate) chordates: Cephalochordate, Tunicata and *Haikouella*. These invertebrates are close relatives of the vertebrates.
- **Non-craniate chordates are close relatives of vertebrates**

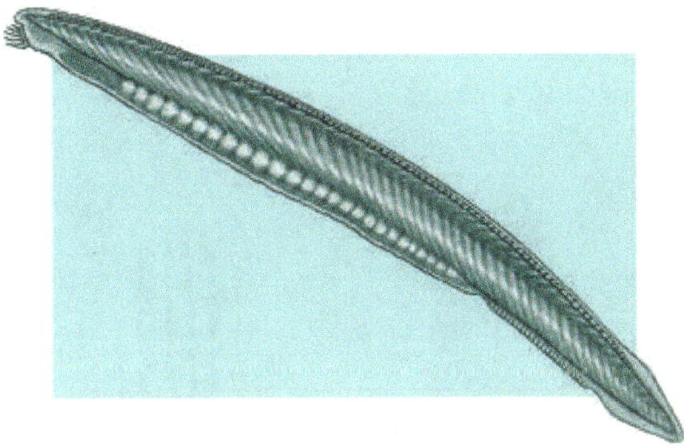

The lancelet, a small translucent fish-like Cephalochordate, is the closest living invertebrate relative of the vertebrates.

Fluorescent-colored sea squirts, *Rhopalaea crassa*. Tunicates may provide clues to vertebrate (and therefore human) ancestry.

Salp chain

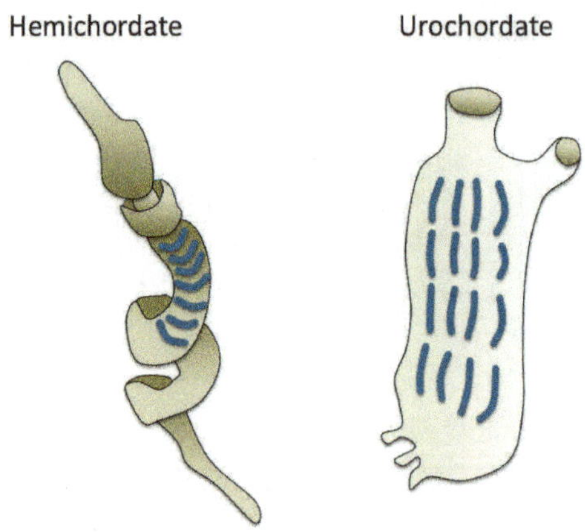

Gill slits in an acorn worm (left) and tunicate (right)

MINERALS FROM SEA WATER

There are a number of marine invertebrates that use minerals that are present in the sea in such minute quantities that they were undetectable until the advent of spectroscopy. Vanadium is concentrated by some tunicates for use in their blood cells to a level ten million times that of the surrounding seawater. Other tunicates similarly concentrate niobium and tantalum. Lobsters use copper in their respiratory pigment hemocyanin, despite the proportion of this metal in seawater being minute. Although these elements are present in vast quantities in the ocean, their extraction by man is not economic.

REFERENCES

1. Gould, Stephen Jay (1990) Wonderful Life: The Burgess Shale and the Nature of History W. W. Norton. ISBN 9780393307009.

2. Davidson, Michael W. (May 26, 2005). Animal Cell Structure. Molecular Expressions. Tallahassee, FL: Florida State University. Retrieved 2008-09-03.
3. Jun-Yuan Chen; Oliveri, Paola; Feng Gao; et al. (August 1, 2002). Precambrian Animal Life: Probable Developmental and Adult Cnidarian Forms from Southwest China (PDF). Developmental Biology. Amsterdam, the Netherlands: Elsevier. 248 (1): 182–196. doi:10.1006/dbio.2002.0714. ISSN 0012-1606. PMID 12142030. Retrieved 2015-02-04.
4. Grazhdankin, Dima (June 2004). Patterns of distribution in the Ediacaran biotas: facies versus biogeography and evolution. Paleobiology. Paleontological Society. 30 (2): 203–221. doi:10.1666/0094-8373(2004)030<0203:PODITE>2.0.CO;2. ISSN 0094-8373.
5. Seilacher, Adolf (August 1992). Vendobionta and Psammocorallia: lost constructions of Precambrian evolution. Journal of the Geological Society. London: Geological Society of London. 149 (4): 607–613. doi:10.1144/gsjgs.149.4.0607. ISSN 0016-7649. Retrieved 2015-02-04.
6. Martin, Mark W.; Grazhdankin, Dmitriy V.; Bowring, Samuel A.; et al. (May 5, 2000). Age of Neoproterozoic Bilaterian Body and Trace Fossils, White Sea, Russia: Implications for Metazoan Evolution. Science. Washington, D.C.: American Association for the Advancement of Science. 288 (5467): 841–845. Bibcode:2000Sci...288..841M. doi:10.1126/science.288.5467.841. ISSN 0036-8075. PMID 10797002. Retrieved 2015-02-05.
7. Fedonkin, Mikhail A.; Waggoner, Benjamin M. (August 28, 1997). The late Precambrian fossil Kimberella is a mollusc-like bilaterian organism. Nature. London: Nature Publishing Group. 388 (6645): 868–871. Bibcode:1997Natur.388..868F. doi:10.1038/42242. ISSN 0028-0836. Retrieved 2008-07-03.
8. Mooi, Rich; David, Bruno (December 1998). Evolution Within a Bizarre Phylum: Homologies of the First Echinoderms (PDF). American Zoologist. Society for Integrative and Comparative Biology. 38 (6): 965–974. doi:10.1093/icb/38.6.965. ISSN 1540-7063. Retrieved 2015-02-05.
9. McMenamin, Mark A. S. (September 2003). Spriggina is a trilobitoid ecdysozoan. Geoscience Horizons Seattle 2003. Abstracts with Programs. 35. Boulder, CO: Geological Society of America. p. 105. OCLC 249088612. Retrieved 2007-11-24. Paper No. 40-2 presented at the Geological Society of America's 2003 Seattle Annual Meeting (November 2–5, 2003) on November 2, 2003, at the Washington State Convention Center.
10. Jih-Pai Lin; Gon, Samuel M., III; Gehling, James G.; et al. (2006). A Parvancorina-like arthropod from the Cambrian of South China. Historical Biology: An International Journal of Paleobiology. Taylor & Francis. 18 (1): 33–45. doi:10.1080/08912960500508689. ISSN 1029-2381.
11. Butterfield, Nicholas J. (December 2006). Hooking some stem-group ‹worms›: fossil lophotrochozoans in the Burgess Shale. BioEssays. Hoboken, NJ: John Wiley & Sons. 28 (12): 1161–1166. doi:10.1002/bies.20507. ISSN 0265-9247. PMID 17120226.

12. Bengtson 2004, pp. 67–78
13. Budd, Graham E. (February 2003). The Cambrian Fossil Record and the Origin of the Phyla (PDF). Integrative and Comparative Biology. Oxford: Oxford University Press for the Society for Integrative and Comparative Biology. 43 (1): 157–165. doi:10.1093/icb/43.1.157. ISSN 1557-7023. PMID 21680420. Retrieved 2015-02-06.
14. Budd, Graham E. (March 1996). The morphology of Opabinia regalis and the reconstruction of the arthropod stem-group. Lethaia. Hoboken, NJ: Wiley-Blackwell. 29 (1): 1–14. doi:10.1111/j.1502-3931.1996.tb01831.x. ISSN 0024-1164.
15. Marshall, Charles R. (May 2006). Explaining the Cambrian ‹Explosion› of Animals. Annual Review of Earth and Planetary Sciences. Palo Alto, CA: Annual Reviews. 34: 355–384. Bibcode:2006AREPS..34..355M. doi:10.1146/annurev.earth.33.031504.103001. ISSN 1545-4495. Retrieved 2015-02-06.
16. Valentine, James W. (2004). On the Origin of Phyla. Chicago: University Of Chicago Press. p. 7. ISBN 0-226-84548-6. Classifications of organisms in hierarchical systems were in use by the seventeenth and eighteenth centuries. Usually organisms were grouped according to their morphological similarities as perceived by those early workers, and those groups were then grouped according to their similarities, and so on, to form a hierarchy.
17. Valentine, James W (2004-06-18). On the Origin of Phyla. ISBN 9780226845487.
18. Erwin, Douglas; Valentine, James; Jablonski, David (1997). Recent fossil finds and new insights into animal development are providing fresh perspectives on the riddle of the explosion of animals during the Early Cambrian. American Scientist (March–April).
19. Budd, G.E.; Jensen, S. (May 2000). A critical reappraisal of the fossil record of the bilaterian phyla. Biological Reviews. 75 (2): 253–295. doi:10.1111/j.1469-185X.1999.tb00046.x. PMID 10881389.
20. Ponder, W.F.; Lindberg, D.R., eds. (2008). Phylogeny and Evolution of the Mollusca. Berkeley: University of California Press. p. 481. ISBN 978-0-520-25092-5.
21. Porifera (n.) Online Etymology Dictionary. Retrieved 18 August 2016.
22. Vacelet & Duport 2004, pp. 179–190.
23. Spongia Linnaeus, 1759. World Register of Marine Species. Retrieved July 18, 2012.
24. Rowland, S. M. & Stephens, T. (2001). Archaeocyatha: A history of phylogenetic interpretation. Journal of Paleontology. 75 (6): 1065–1078. doi:10.1666/0022-3360(2001)075<1065:AAHOPI>2.0.CO;2. JSTOR 1307076. Archived from the original on December 6, 2008.
25. Sperling, E. A.; Pisani, D.; Peterson, K. J. (January 1, 2007). Poriferan paraphyly and its implications for Precambrian palaeobiology (PDF). Geological Society,

London, Special Publications. 286 (1): 355–368. doi:10.1144/SP286.25. Retrieved August 22, 2012.

26. Zhang, Z.-Q. (2011). Animal biodiversity: An introduction to higher-level classification and taxonomic richness (PDF). Zootaxa. 3148: 7–12.
27. Nematostella vectensis v1.0. Genome Portal. University of California. Retrieved 2014-01-19.
28. Nematostella. Nematostella.org. Retrieved 2014-01-18.
29. Genikhovich, G. & U. Technau (2009). The Starlet sea anemone Nematostella vectensis: An anthozoan model organism for studies in comparative genomics and functional evolutionary developmental biology. Cold Spring Harbor Protocols. 2009. doi:10.1101/pdb.emo129.
30. Where Does Our Head Come From? Brainless Sea Anemone Sheds New Light on the Evolutionary Origin of the Head. Science Daily. 2013-02-12. Retrieved 2014-01-18.
31. Sinigaglia, C.; et al. (2013). The bilaterian head patterning gene six3/6 controls aboral domain development in a cnidarian. PLoS Biology. 11 (2): e1001488. doi:10.1371/journal.pbio.1001488.
32. Bavestrello, Giorgio; Christian Sommer; Michele Sarà (1992). Bi-directional conversion in Turritopsis nutricula (Hydrozoa). Scientia Marina. 56 (2–3): 137–140.
33. Piraino, Stefano; F. Boero; B. Aeschbach; V. Schmid (1996). Reversing the life cycle: medusae transforming into polyps and cell transdifferentiation in Turritopsis nutricula (Cnidaria, Hydrozoa). Biological Bulletin. Biological Bulletin, vol. 190, no. 3. 190 (3): 302–312. doi:10.2307/1543022. JSTOR 1543022.
34. Aurelia aurita Encyclopedia of Life. Retrieved 18 August 2016.
35. Karleskint G, Richard Turner R and , James Small J (2012) Introduction to Marine Biology Cengage Learning, edition 4, page 445. ISBN 9781133364467.
36. Cornwall – Nature – Superstar Worm. BBC.
37. Mark Carwardine (1995) The Guinness Book of Animal Records. Guinness Publishing. p. 232.
38. The Persistent Parasites. Time Magazine. Time Inc. 1957-04-08.
39. Hargis, William J. (1985). Parasitology and pathology of marine organisms of the world ocean. NOAA Tech. Rep. National Oceanic and Atmospheric Administration.
40. http://plpnemweb.ucdavis.edu/nemaplex/General/animpara.htm
41. https://cid.oxfordjournals.org/content/29/4/734.full.pdf
42. Hodda, M (2011). Phylum Nematoda Cobb, 1932. In: Zhang, Z.-Q. (Ed.) Animal biodiversity: An outline of higher-level classification and survey of taxonomic richness. Zootaxa. 3148: 63–95.
43. Zhang, Z (2013). Animal biodiversity: An update of classification and diversity in 2013. In: Zhang, Z.-Q. (Ed.) Animal Biodiversity: An Outline of Higher-level

Classification and Survey of Taxonomic Richness (Addenda 2013). Zootaxa. 3703 (1): 5–11. doi:10.11646/zootaxa.3703.1.3.

44. Lambshead PJD (1993). Recent developments in marine benthic biodiversity research. Oceanis. 19 (6): 5–24.
45. Borgonie G, García-Moyano A, Litthauer D, Bert W, Bester A, van Heerden E, Möller C, Erasmus M, Onstott TC (June 2011). Nematoda from the terrestrial deep subsurface of South Africa. Nature. 474 (7349): 79–82. doi:10.1038/nature09974. PMID 21637257.
46. Danovaro R, Gambi C, Dell›Anno A, Corinaldesi C, Fraschetti S, Vanreusel A, Vincx M, Gooday AJ (January 2008). Exponential decline of deep-sea ecosystem functioning linked to benthic biodiversity loss. Curr. Biol. 18 (1): 1–8. doi:10.1016/j.cub.2007.11.056. PMID 18164201. Lay summary – EurekAlert!.
47. Platt HM (1994). foreword. In Lorenzen S, Lorenzen SA. The phylogenetic systematics of freeliving nematodes. London: The Ray Society. ISBN 0-903874-22-9.
48. Sea Lily. Science Encyclopedia. Retrieved Sep 5, 2014.
49. Animal Diversity Web - Echinodermata. University of Michigan Museum of Zoology. Retrieved 26 August 2012.
50. Fox, Richard. Asterias forbesi. Invertebrate Anatomy OnLine. Lander University. Retrieved 14 June 2014.
51. Holsinger, K. (2005). Keystone species. Retrieved May 10, 2010, from http://darwin.eeb.uconn.edu/eeb310/lecture-notes/interactions/node2.html
52. Chapman, A.D. (2009). Numbers of Living Species in Australia and the World, 2nd edition. Australian Biological Resources Study, Canberra. Retrieved 12 January 2010. ISBN 978-0-642-56860-1 (printed); ISBN 978-0-642-56861-8 (online).
53. Hancock, Rebecca (2008). Recognising research on molluscs. Australian Museum. Retrieved 2009-03-09.
54. Barnes, R.S.K., Calow, P., Olive, P.J.W., Golding, D.W. and Spicer, J.I. (2001). The Invertebrates, A Synthesis (3 ed.). UK: Blackwell Science.
55. Welcome to CephBase. CephBase. Retrieved 29 January 2016.
56. Wilbur, Karl M.; Clarke, M.R.; Trueman, E.R., eds. (1985), The Mollusca, 12. Paleontology and neontology of Cephalopods, New York: Academic Press, ISBN 0-12-728702-7
57. Black, Richard (April 26, 2008). Colossal squid out of the freezer. BBC News.
58. Ewen Callaway (2 June 2008). Simple-Minded Nautilus Shows Flash of Memory. New Scientist. Retrieved 7 March 2012.
59. Kathryn Phillips (15 June 2008). Living Fossil Memories (PDF). Inside JEB. 211 (12): iii. doi:10.1242/jeb.020370.
60. Robyn Crook & Jennifer Basil (2008). A biphasic memory curve in the chambered nautilus, Nautilus pompilius L. (Cephalopoda: Nautiloidea) (PDF). The Journal of Experimental Biology. 211 (12): 1992–1998. doi:10.1242/jeb.018531.
61. Ruppert, Edward E.; Fox, Richard, S.; Barnes, Robert D. (2004). Invertebrate Zoology, 7th edition. Cengage Learning. ISBN 81-315-0104-3.

62. Hayward, PJ (1996). Handbook of the Marine Fauna of North-West Europe. Oxford University Press. ISBN 0-19-854055-8.

63. Giribet, G.; Okusu, A, A.; Lindgren, A.R., A. R.; Huff, S.W., S. W.; Schrödl, M, M.; Nishiguchi, M.K., M. K. (May 2006). Evidence for a clade composed of molluscs with serially repeated structures: monoplacophorans are related to chitons (Free full text). Proceedings of the National Academy of Sciences of the United States of America. 103 (20): 7723–7728. Bibcode:2006PNAS..103.7723G. doi:10.1073/pnas.0602578103. PMC 1472512. PMID 16675549.

64. Healy, J.M. (2001). The Mollusca. In Anderson, D.T. Invertebrate Zoology (2 ed.). Oxford University Press. pp. 120–171. ISBN 0-19-551368-1.

65. http://firstlifeseries.com/learn-more/

66. D. R. Currie & T. M. Ward (2009). South Australian Giant Crab (Pseudocarcinus gigas) Fishery (PDF). South Australian Research and Development Institute. Fishery Assessment Report for PIRSA. Retrieved 9 December 2013.

67. Gewin, V. (2005). "Functional genomics thickens the biological plot". PLoS biology, 3 (6), e219. doi:10.1371/journal.pbio.0030219

68. Lancelet (amphioxus) genome and the origin of vertebrates Ars Technica, 19 June 2008.

69. Lemaire, P. (2011) "Evolutionary crossroads in developmental biology: the tunicates". Development, 138 (11): 2143–2152. doi:10.1242/dev.048975

70. Carson, Rachel (1997). The Sea Around Us. Oxford Paperbacks. pp. 190–191. ISBN 0195069978.

CHAPTER 5

Marine Fungi

Chapter Contents

Marine fungi are distinct from terrestrial and freshwater fungi in their taxonomy and morphology, as well as in their adaptations to an aquatic habitat. They are an ecological, not a taxonomic group and cannot be defined by nutritional or physiological requirements. Obligate marine fungi are those that grow and sporulate exclusively in a marine or estuarine habitat and are permanently or intermittently submerged. Facultative marine fungi are those that normally occupy freshwater habitats or terrestrial milieus but are able to grow (and possibly to sporulate) in the marine environment.

INTRODUCTION

Marine fungi are species of fungi that live in marine or estuarine environments. They are not a taxonomic group but share a common habitat. Obligate marine fungi grow exclusively in the marine habitat while wholly or sporadically submerged in sea water. Facultative marine fungi normally occupy terrestrial or freshwater habitats but are capable of living or even sporulating in a marine habitat. About 444 species of marine fungi have been described, including 7 genera and ten species of basidiomycetes, and 177 genera and 360 species of ascomycetes. The remainder of the marine fungi are chytrids and mitosporic or asexual fungi. Many species of marine fungi are known only from spores and it is likely a large number of species have yet to be discovered. It is impracticable to culture many of these fungi but their nature can be investigated by examining sea water samples and undertaking rDNA analysis of the fungal material found.

Different marine habitats support very different fungal communities. Fungi can be found in niches ranging from ocean depths and coastal waters to mangrove swamps and estuaries with low salinity levels. Marine fungi can be saprobic or parasitic on animals, saprobic or parasitic on algae, saprobic on plants or saprobic on dead wood.

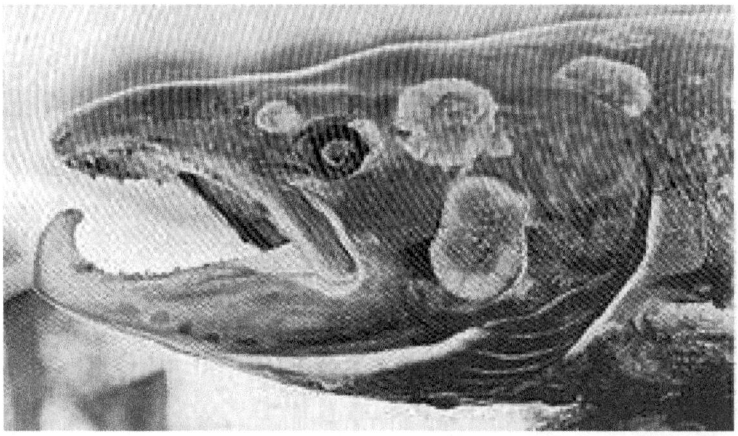

Mature salmon with fungal disease

TYPES OF MARINE FUNGI

Factors that influence whether or not marine fungi are present in any particular location include the water temperature, its salinity, the water movement, the presence of suitable substrates for colonization, the presence of propagules in the water, interspecific competition, pollution and the oxygen content of the water.

Some marine fungi which have ventured into the sea from terrestrial habitats include species that burrow into sand grains, living in the pores. Others live inside stony corals, and may become pathogenic if the coral is stressed by rising sea temperatures.

In 2011 the phylogeny of marine fungi was elucidated by analysis of their small subunit ribosomal DNA sequences. Thirty six new marine lineages were found, the majority of which were chytrids but also some filamentous and multicellular fungi. The majority of the species found were ascomycetous and basidiomycetous yeasts.

The secondary metabolites produced by marine fungi have high potential for use in biotechnological, medical and industrial applications.

MANGROVES

Mangroves in Puerto Rico

The greatest number of known species of marine fungi are from mangrove swamps. In one study, blocks of mangrove timber and pieces of driftwood of *Avicennia alba*, *Bruguiera cylindrica* and *Rhizophora apiculata* were examined to identify the lignicolous (wood-decaying) fungi they hosted. Also tested were *Nypa fruticans*, a mangrove palm and *Acanthus ilicifolius*, a plant often associated with mangroves. Each material was found to have its own characteristic fungi, the greatest diversity being among those growing on the mangrove palm. It was surmised that this was because the salinity was lower in the estuaries and creeks where *Nypa* grew, and so it required a lesser degree of adaptation for the fungi to flourish there. Some of these species were closely related to fungi on terrestrial palms. Other studies have shown that driftwood hosts more species of fungus than do exposed test blocks of timber of a similar kind. The mangrove leaf litter also supported a large fungal community which was different from that on the timber and living material. However, few of these were multicellular, higher marine fungi.

OTHER PLANTS

The sea snail *Littoraria irrorata* damages plants of *Spartina* in the sea marshes where it lives, which enables spores of intertidal ascomycetous fungi to colonise the plant. The snail eats the fungal growth in preference to the grass itself. This mutualism between the snail and the fungus is considered to be the first example of husbandry among invertebrate animals outside the class Insecta.

Eelgrass, *Zostera marina*, is sometimes affected by seagrass wasting disease. The primary cause of this seems to be pathogenic strains of the protist, *Labyrinthula zosterae*, but it is thought that fungal pathogens also contribute and may predispose the eelgrass to disease.

Timber

Driftwood on Jekyll Island, Georgia

Many marine fungi are very specific as to which species of floating and submerged timbers they colonise. A range of species of fungi colonise beech while oak supports a different community. When a fungal propagule lands on a suitable piece of timber, it will grow if no other fungi are present. If the log is already colonised by another fungal species, growth will depend on whether that fungus produces antifungal chemicals and whether the new arrival can resist them. The chemical properties of colonizing fungi also affect the animal communities that graze on them: in one study, when hyphae from five different species of marine fungi were fed to nematodes, one species supported less than half the number of nematodes per mg of hyphae than did the others. Detection of fungi in timber may involve incubation at a suitable temperature in a suitable water medium for a period of six months to upward of eighteen months.

ALGAE

Rhyzophydium littoreum is a marine chytrid, a primitive fungus that infects green algae in estuaries. It obtains nutrients from the host alga and produces swimming zoospores that must survive in open water, a low nutrient environment, until a new host is encountered.

A species of *Fusarium* isolated from a seaweed has been shown to contain a chemical that seems to have cytogenic, anti-cancer effects. Another fungus, *Ascochyta salicorniae*, found growing on seaweed is being investigated for its action against malaria, a mosquito-borne infectious disease of humans and other animals.

LICHENS

Lichens are mutualistic associations between a fungus, usually an ascomycete, and an alga or a cyanobacterium. Several lichens, including *Arthopyrenia halodytes*, *Pharcidia laminariicola*, *Pharcidia rhachiana* and *Turgidosculum ulvae*, are found in marine environments. Many more occur in the splash zone, where they occupy different vertical zones depending on how tolerant they are to submersion. Fossil marine lichens 600 million years old have been discovered in the late Neoproterozoic marine phosphate rocks in the sedimentary, fossil-rich Doushantuo Formation in China.

Lichen covered rocks on the Isles of Scilly

VERTEBRATES

Whales, porpoises and dolphins are susceptible to fungal diseases but these have been little researched in the field. Mortalities from fungal disease have been reported in captive killer whales; it is thought that stress due to captive conditions may have been predisposing. Transmission among animals in the open sea may naturally limit the spread of fungal diseases. Infectious fungi known from killer whales include *Aspergillus fumigatus*, *Candida albicans* and *Saksenaea vasiformis*. Fungal infections in other cetaceans include *Coccidioides immitis*, *Cryptococcus neoformans*, *Loboa loboi*, *Rhizopus* sp., *Aspergillus flavus*, *Blastomyces dermatitidus*, *Cladophialophora bantiana*, *Histoplasma capsulatum*, *Mucor* sp., *Sporothrix schenckii* and *Trichophyton* sp.

Salmonids farmed in cages in marine environments may be affected by a number of different fungal infections. *Exophiala salmonis* causes an infection in which growth of hyphae in the kidneys causes swelling of the abdomen. A cellular response by the fish aims to isolate the fungus by walling it off. Fish are also susceptible to fungus-like oomycetes including *Branchiomyces* which affects the gills of various fishes, and*Saprolegnia* which attacks damaged tissue.

INVERTEBRATES

The American lobster *(Homarus americanus)*, like many other marine crustaceans, incubates its eggs beneath its tail segments. Here they are exposed to water-borne micro-organisms including fungi during their long period of development. The lobster has a symbiotic relationship with a gram-negative bacterium that has anti-fungal properties. This bacterium grows over the eggs and protects them from infection by the pathogenic fungus-like oomycete *Lagenidium callinectes*. The metabolite produced by the bacterium is tyrosol, a 4-hydroxyphenethyl alcohol, an antibiotic substance also produced by some terrestrial fungi. Similarly, a shrimp found in estuaries, *Palaemon macrodactylis*, has a symbiotic bacterium that produces 2,3-indolenedione, a substance that is also toxic to the oomycete *Lagenidium callinectes*.

REFERENCES

1. Species of Higher Marine Fungi University of Mississippi. Retrieved 2012-02-05.
2. Cool New Paper: Marine Fungi Teaching Biology. Retrieved 2012-02-05.
3. E. B. Gareth Jones (2000). Marine fungi: some factors influencing biodiversity (PDF). Fungal Diversity. 4: 53–73.
4. Holmquist, G. U.; H. W. Walker & Stahr H. M. (1983). Influence of Temperature, pH, Water Activity and Antifungal Agents on Growth of Aspergillus flavus and A. parasiticus. Journal of Food Science. 48: 778–782. doi:10.1111/j.1365-2621.1983.tb14897.x.
5. Richards, Thomas A.; Meredith D.M. Jones; Guy Leonard; David Bass (2011). Marine Fungi: Their Ecology and Molecular Diversity. Annual Review of Marine Science. doi:10.1146/annurev-marine-120710-100802.
6. Marine Fungi Retrieved 2012-02-06.
7. Silliman B. R. & S. Y. Newell (2003). Fungal farming in a snail. PNAS. 100 (26): 15643–15648. doi:10.1073/pnas.2535227100. PMC 307621. PMID 14657360.
8. Disease Analysis in San Juan Archipelago Friday Harbor Laboratories Seagrass Lab. Retrieved 2012-02-06.
9. Short, Frederick T.; Robert G. Coles. Global seagrass research methods. Google Books. p. 414.
10. Moss, Stephen T. (1986). The biology of marine fungi. Google Books. pp. 65–70.
11. Ebel R. (2006). Secondary metabolites from marine derived fungi. Frontiers in Marine Biotechnology.
12. Osterhage C.; R. Kaminsky; G. König & A. D. Wright (2000). Ascosalipyrrolidinone A, an Antimicrobial Alkaloid, from the Obligate Marine Fungus Ascochyta salicorniae. Journal of Organic Chemistry. 65 (20): 6412–6417. doi:10.1021/jo000307g. PMID 11052082.
13. Freshwater and marine lichen-forming fungi Retrieved 2012-02-06.
14. Yuan X, Xiao S & Taylor TN. (2005). Lichen-Like Symbiosis 600 Million Years Ago. Science. 308: 1017–1020. doi:10.1126/science.1111347. PMID 15890881.
15. Joseph K. Gaydos, JK; KC Balcomb; RW Osborn; L Dierauf. A Review of Potential Infectious Disease Threats to Southern Resident Killer Whales (Orcinus orca) (PDF).
16. Fungal infections of farmed salmon and trout Retrieved 2012-02-06.
17. Gil-Turnes, M. Sofia & William Fenical (1992). Embryos of Homarus americanus are Protected by Epibiotic Bacteria. The Biological Bulletin. 182: 105–108. JSTOR 1542184.

CHAPTER 6

Marine Vertebrate

Chapter Contents

The Marine Vertebrate Biology cluster is one of the largest in the institute and includes researchers who study a broad range of marine mammals and birds including dolphins, whales, sea otters, seals, sea lions, peregrine falcons, eagles, and a variety of marine birds. This area of research was one of the first established within the Institute and the scientists in this group have had great success and are internationally recognized for developing unique field and lab techniques to investigate wild populations of marine mammals, marine birds, and raptors.

INTRODUCTION

Marine vertebrates are vertebrates which live in a marine environment. These primarily include fish, seabirds, marine reptiles, and marine mammals. These animals have an internal skeleton and make up about 4% of the sea`s animal population.

MARINE ANIMALS: VERTEBRATES

Amphibians mostly require fresh water to reproduce. A few inhabit brackish water, but there are no true marine amphibians. There have been reports, however, of amphibians invading marine waters, such as a Black Sea invasion by the natural hybrid *Pelophylax esculentus* reported in 2010.

Fish anatomy includes a two-chambered heart, operculum, swim bladder, scales, eyes adapted to seeing underwater, and secretory cells that produce mucous. Fish breathe by extracting oxygen from water through gills. Fins propel and stabilize the fish in the water. Fish fall into two main groups: fish with bony skeletons and fish with cartilaginous skeletons.

A reported 32,700 species of fish have been described (as of December 2013), more than the combined total of all other vertebrates. About 60% of fish species are saltwater fish.

AGNATHA

Agnatha (Greek, "no jaws") is a superclass of jawless fish in the phylum Chordata, subphylum Vertebrata, consisting of both present (cyclostomes) and extinct (conodonts and ostracoderms) species. The group excludes all vertebrates with jaws, known as gnathostomes.

The agnathans as a whole are paraphyletic, because most extinct agnathans belong to the stem group of gnathostomes. Recent molecular data, both from rRNA and from mtDNA as well as embryological data strongly supports the hypothesis that living agnathans, the cyclostomes, are monophyletic.

The oldest fossil agnathans appeared in the Cambrian, and two groups still survive today: the lampreys and the hagfish, comprising about 120 species in total. Hagfish are considered members of the subphylum Vertebrata, because they secondarily lost vertebrae; before this event was inferred from molecular and developmentaldata, the group Craniata was created by Linnaeus (and is still sometimes used as a strictly morphological descriptor) to reference hagfish plus vertebrates. In addition to the absence of jaws, modern agnathans are characterised by absence of paired fins; the presence of a notochord both in larvae and adults; and seven or more paired gill pouches. Lampreys have a light sensitive pineal eye (homologous to the pineal gland in mammals). All living and most extinct Agnatha do not have an identifiable stomach or any appendages. Fertilization and development are both external. There is no parental care in the Agnatha class. The Agnatha are ectothermic or cold blooded, with a cartilaginous skeleton, and the heart contains two chambers.

While a few scientists still regard the living agnathans as only superficially similar, and argue that many of these similarities are probably shared basal characteristics of ancient vertebrates, recent classifications clearly place hagfish (the Myxini or Hyperotreti), with the lampreys (Hyperoartii) as being more closely related to each other than either is to the jawed fishes.

Metabolism

Agnathans are ectothermic, meaning they do not regulate their own body temperature. Agnathan metabolism is slow in cold water, and therefore they do not have to eat very much. They have no distinct stomach, but rather a long gut, more or less homogenous throughout its length. Lampreys feed on other fish and mammals. They rely on a row of sharp teeth to shred their host. Anticoagulant fluids preventing blood clotting are injected into the host, causing the host to yield more blood. Hagfish are scavengers, eating mostly dead animals. They also use a sharp set of teeth to break down the animal. The fact that Agnathan teeth are unable to move up and down limits their possible food types.

Body Covering

In modern agnathans, the body is covered in skin, with neither dermal or epidermal scales. The skin of hagfish has copious slime glands, the slime constituting their defense mechanism. The slime can sometimes clog up enemy fishes' gills, causing them to die. In direct contrast, many extinct agnathans sported extensive exoskeletons composed of either massive, heavy dermal armour or small mineralized scales.

Appendages

Almost all agnathans, including all extant agnathans, have no paired appendages, although most do have a dorsal or a caudal fin. Some fossil agnathans, such as osteostracans and pituriaspids, did have paired fins, a trait inherited in their jawed descendants.

Reproduction

Fertilization in lampreys is external. Mode of fertilization in hagfishes is not known. Development in both groups probably is external. There is no known parental care. Not much is known about the hagfish reproductive process. It is believed that hagfish only have 30 eggs over a lifetime. Most species are hermaphrodites. There is very little of the larval stage that characterizes the lamprey. Lamprey are only able to reproduce once. After external fertilization, the lamprey's cloacas remain open, allowing a fungus to enter their intestines, killing them. Lampreys reproduce in freshwater riverbeds, working in pairs to build a nest and burying their eggs about an inch beneath the sediment. The resulting hatchlings go through four years of larval development before becoming adults. They also have a certain unusual form of reproduction.

Evolution

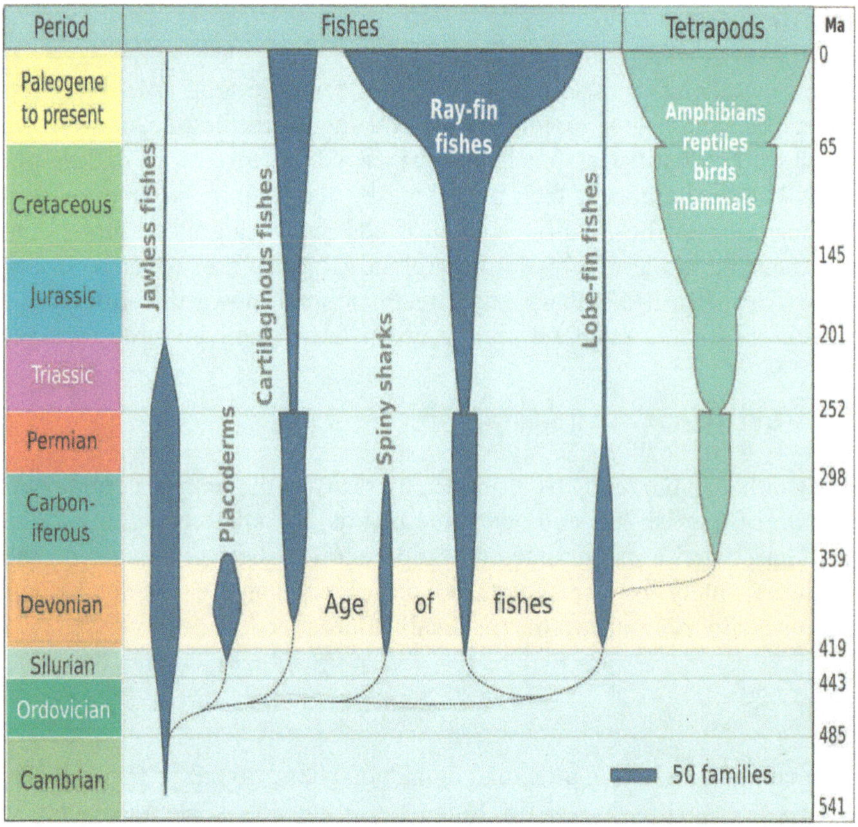

Evolution of jawless fishes. The diagram is based on Michael Benton, 2005.

Although a minor element of modern marine fauna, agnathans were prominent among the early fish in the early Paleozoic. Two types of Early Cambrian animal apparently having fins, vertebrate musculature, and gills are known from the early Cambrian Maotianshan shales of China: Haikouichthys and Myllokunmingia. They have been tentatively assigned to Agnatha by Janvier. A third possible agnathid from the same region is Haikouella. A possible agnathid that has not been formally described was reported by Simonetti from the Middle Cambrian Burgess Shale of British Columbia.

Many Ordovician, Silurian, and Devonian agnathans were armored with heavy bony-spiky plates. The first armored agnathans—the Ostracoderms, precursors to the bony fish and hence to the tetrapods (including humans)—are known from the middle Ordovician, and by the Late Silurian the agnathans had reached the high point of their evolution. Most of the ostracoderms, such as thelodonts, osteostracans, and galeaspids, were more closely related to the gnathostomes than to the surviving agnathans, known as cyclostomes. Cyclostomes apparently split from other agnathans before the evolution of dentine and bone, which are present in many fossil agnathans, including conodonts. Agnathans declined in the Devonian and never recovered.

Classification

Subgroups of jawless fish			
Subgroup		**Example**	**Comments**
Cyclostomes	**Myxini**	 hagfish	Myxini (hagfish) are eel-shaped slime-producing marine animals (occasionally called slime eels). They are the only known living animals that have a skull but not a vertebral column. Along with lampreys, hagfish are jawless and are living fossils; hagfish are basal to vertebrates, and living hagfish remain similar to hagfish 300 million years ago.The classification of hagfish has been controversial. The issue is whether the hagfish is itself a degenerate type of vertebrate-fish (most closely related to lampreys), or else may represent a stage which precedes the evolution of the vertebral column (as do lancelets). The original scheme groups hagfish and lampreys together as cyclostomes (or historically, Agnatha), as the oldest surviving clade of vertebrates alongside gnathostomes (the now-ubiquitous jawed-vertebrates). An alternative scheme proposed that jawed-vertebrates are more closely related to lampreys than to hagfish (i.e., that vertebrates include lampreys but exclude hagfish), and introduces the category craniata to group vertebrates near hagfish. Recent DNA evidence has supported the original scheme.
	Hyperoartia	 lamprey	Hyperoartia is a disputed group of vertebrates that includes the modern lampreys and their fossil relatives. Examples of hyperoartians from early in their fossil record are Endeiolepis and Euphanerops, fish-like animals with hypocercal tails that lived during the Late Devonian Period. Some paleontologists still place these forms among the "ostracoderms" (jawless armored "fishes") of the classAnaspida, but this is increasingly considered an artificial arrangement based on ancestral traits. Placement of this group among the jawless vertebrates is a matter of dispute. While today enough fossil diversity is known to make a close relationship among the "ostracoderms" unlikely, this has muddied the issue of the Hyperoartia's closest relatives. Traditionally the group was placed in a superclass Cyclostomata together with the Myxini (hagfishes). More recently, it has been proposed that the Myxini are more basal among the skull-bearing chordates, while the Hyperoartia are retained among vertebrates. But even though this may be correct, the lampreys represent one of the oldest divergences of the vertebrate lineage, and whether they are better united with some "ostracoderms" in the Cephalaspidomorphi, or not closer to these than to e.g. to other "ostracoderms" of the Pteraspidomorphi, or even the long-extinct conodonts, is still to be resolved. Even the very existence of the Hyperoartia is disputed, with some analyses favoring a treatment of the "basal Hyperoartia" as a monophyletic lineage Jamoytiiformes that may in fact be very close to the ancestral jawed vertebrates.

<table>
<tr>
<td rowspan="4">Ostracoderms</td>
<td>†Pteraspidomorphi (extinct)</td>
<td></td>
<td>†Pteraspidomorphi is an extinct group of early jawless fish. The fossils show extensive shielding of the head. Many had hypocercal tails in order to generate lift to increase ease of movement through the water for their armoured bodies, which were covered in dermal bone. They also had sucking mouth parts and some species may have lived in fresh water.

The taxon contains the subgroups Heterostraci, Astraspida, Arandaspida.</td>
</tr>
<tr>
<td>†Thelodonti (extinct)</td>
<td></td>
<td>Thelodonti (nipple teeth) are a group of small, extinct jawless fishes with distinctive scales instead of large plates of armour. There is much debate over whether the group of Palaeozoic fish known as the Thelodonti (formerly coelolepids) represent a monophyletic grouping, or disparate stem groups to the major lines of jawless and jawed fish. Thelodonts are united in possession of "thelodont scales". This defining character is not necessarily a result of shared ancestry, as it may have been evolved independently by different groups. Thus the thelodonts are generally thought to represent a polyphyletic group,although there is no firm agreement on this point; if they are monophyletic, there is no firm evidence on what their ancestral state was. "Thelodonts" were morphologically very similar, and probably closely related, to fish of the classes Heterostraci and Anaspida, differing mainly in their covering of distinctive, small, spiny scales. These scales were easily dispersed after death; their small size and resilience makes them the most common vertebrate fossil of their time. The fish lived in both freshwater and marine environments, first appearing during the Ordovician, and perishing during the Frasnian–Famennian extinction event of the Late Devonian. They were predominantly deposit-feeding bottom dwellers, although there is evidence to suggest that some species took to the water column to be free-swimming organisms.</td>
</tr>
<tr>
<td>†Anaspida (extinct)</td>
<td></td>
<td>Anaspida (without shield) is an extinct group of primitive jawless vertebrates that lived during the Silurian and Devonian periods. They are classically regarded as the ancestors of lampreys. Anaspids were small marine agnathans that lacked heavy bony shield and paired fins, but have a striking highly hypocercal tail. They first appeared in the early Silurian, and flourished until the Late Devonian extinction, where most species, save for lampreys, became extinct due to the environmental upheaval during that time.</td>
</tr>
<tr>
<td>†Cephalaspido-morphi (extinct)</td>
<td></td>
<td>Cephalaspidomorphi is a broad group of extinct armored agnathans found in Silurian and Devonian strata of North America, Europe, and China, and is named in reference to the osteostracan genus Cephalaspis. Most biologists regard this taxon as extinct, but the name is sometimes used in the classification of lampreys, as lampreys are sometimes thought to be related to cephalaspids. If lampreys are included, they would extend the known range of the group from the early Silurian period through the Mesozoic, and into the present day. Cephalaspidomorphi were, like most contemporary fish, very well armoured. Particularly the head shield was well developed, protecting the head, gills and the anterior section of the innards. The body was in most forms well armoured as well. The head shield had a series of grooves over the whole surface forming an extensive lateral line organ. The eyes were rather small and placed on the top of the head. There was no proper jaw. The mouth opening was surrounded by small plates making the lips flexible, but without any ability to bite. Undisputed subgroups traditionally contained with Cephaloaspidomorphi, also called "Monorhina," include the classes Osteostraci, Galeaspida, and Pituriaspida</td>
</tr>
</table>

Jawed fish

Cartilaginous

The Greenland shark has the longest known lifespan of all vertebrates, about 400 years. Megalodon is an extinct species of shark that lived about 28 to 1.5 Ma. It looked much like a stocky version of the great white shark, but was much larger with fossil lengths reaching 20.3 m (67 ft). Found in all oceans it was one of the largest and most powerful predators in vertebrate history, and probably had a profound impact on marine life.

Cartilaginous fishes may have evolved from spiny sharks

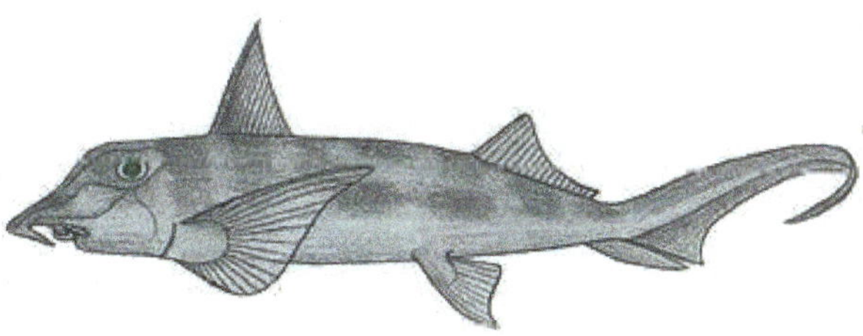

Chimaera

Caribbean reef shark

Megalodon resembled a giant great white shark

Stingray

Sawfish are rays with long rostrums resembling a saw. All are now endangered or critically endangered

The giant manta ray, largest ray in the world, has been targeted by fisheries until it is vulnerable.

Bony

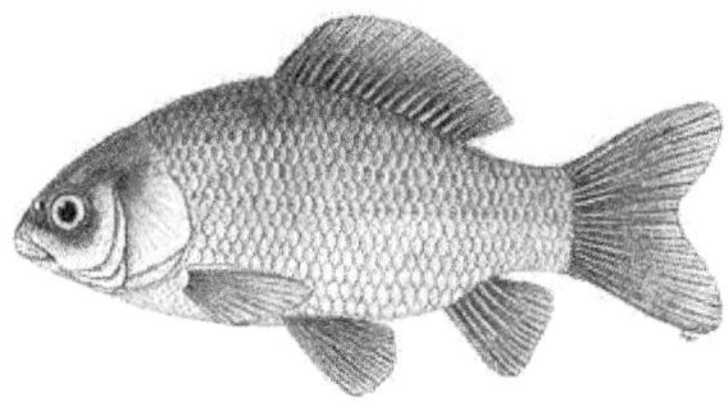

Ray-finned fish (crucian carp)

Lobe-finned fish (coelacanth)

The extinct *Leedsichthys* is the largest known bony fish

Marine Tetrapods

Reptiles

Reptiles (Late Latin for *creeping* or *crawling*) form a class Reptilia of tetrapods, vertebrates that have either four limbs or have four-limbed ancestors. Some reptiles are more closely related to birds than other reptiles, and many scientists prefer to make Reptilia a monophyletic group which includes the birds. Extant non-avian reptiles which inhabit or frequent the sea include sea turtles, sea snakes, terrapins, the marine iguana, and the saltwater crocodile. Currently, of the approximately 12,000 extant reptile species and sub-species, only about 100 of are classed as marine reptiles.

Unlike amphibians, reptiles do not have an aquatic larval stage. Most reptiles are oviparous, although several species of squamates are viviparous, as were some extinct aquatic clades—the fetus develops within the mother, contained in a placenta rather than an eggshell. As amniotes, reptile eggs are surrounded by membranes for protection and transport, which adapt them to reproduction on dry land. Many of the viviparous species feed their fetuses through various forms of placenta analogous to those of mammals, with some providing initial care for their hatchlings.

Except for some sea snakes, most extant marine reptiles are oviparous and need to return to land to lay their eggs. Apart from sea turtles, the species usually spend most of their lives on or near land rather than in the ocean. Sea snakes generally prefer shallow waters nearby land, around islands, especially waters that are somewhat sheltered, as well as near estuaries.Unlike land snakes, sea snakes have evolved flattened tails which help them swim.

Marine iguana

Leatherback sea turtle

Saltwater crocodile

Marine snakes have flattened tails

The ancient *Ichthyosaurus communis* independently evolved flippers similar to dolphins

Some extinct marine reptiles, such as ichthyosaurs, evolved to be viviparous and had no requirement to return to land. Ichthyosaurs resembled dolphins. They first appeared about 245 million years ago and disappeared about 90 million years ago. The terrestrial ancestor of the ichthyosaur had no features already on its back or tail that might have helped along the evolutionary process. Yet the ichthyosaur developed a dorsal and tail fin which improved its ability to swim. The biologist Stephen Jay Gould said the ichthyosaur was his favourite example of convergent evolution. The earliest marine reptiles arose in the Permian. During the Mesozoic many groups of reptiles became adapted to life in the seas, including ichthyosaurs, plesiosaurs, mosasaurs, nothosaurs, placodonts, sea turtles, thalattosaurs and thalattosuchians. Marine reptiles were less numerous after mass extinction at the end of the Cretaceous.

SEABIRDS

Seabirds (also known as **marine birds**) are birds that are adapted to life within the marine environment. While seabirds vary greatly in lifestyle, behaviour and physiology, they often exhibit striking convergent evolution, as the same environmental problems and feeding niches have resulted in similar adaptations. The first seabirds evolved in the Cretaceous period, and modern seabird families emerged in the Paleogene.

The sooty tern is highly aerial and marine and will spend months flying at sea, returning to land only for breeding.

In general, seabirds live longer, breed later and have fewer young than other birds do, but they invest a great deal of time in their young. Most species nest in colonies, which can vary in size from a few dozen birds to millions. Many species are famous for undertaking long annual migrations, crossing the equator or circumnavigating the Earth in some cases. They feed both at the ocean's surface and below it, and even feed on each other. Seabirds can be highly pelagic, coastal, or in some cases spend a part of the year away from the sea entirely.

Seabirds and humans have a long history together: they have provided food to hunters, guided fishermen to fishing stocks and led sailors to land. Many species are currently threatened by human activities, and conservation efforts are under way.

Classification of Seabirds

There exists no single definition of which groups, families and species are seabirds, and most definitions are in some way arbitrary. In the words of two seabird scientists, "The one common characteristic that all seabirds share is that they feed in saltwater; but, as seems to be true with any statement in biology, some do not." However, by convention all of the Sphenisciformes and Procellariiformes, all of the Pelecaniformes except the darters, and some of the Charadriiformes (the skuas, gulls, terns, auks and skimmers) are classified as seabirds. The phalaropes are usually included as well, since although they are waders ("shorebirds" in North America), two of the three species are oceanic for nine months of the year, crossing the equator to feed pelagically.

Loons and grebes, which nest on lakes but winter at sea, are usually categorized as water birds, not seabirds. Although there are a number of sea ducks in the family Anatidae that are truly marine in the winter, by convention they are usually excluded from the seabird grouping. Many waders (or shorebirds) and herons are also highly marine, living on the sea's edge (coast), but are also not treated as seabirds.

Evolution and Fossil Record

Seabirds, by virtue of living in a geologically depositional environment (that is, in the sea where sediments are readily laid down), are well represented in the fossil record. They are first known to occur in the Cretaceous period, the earliest being the Hesperornithiformes, like *Hesperornis regalis*, a flightless loon-like seabird that could dive in a fashion similar to grebes and loons (using its feet to move underwater) but had a beak filled with sharp teeth.

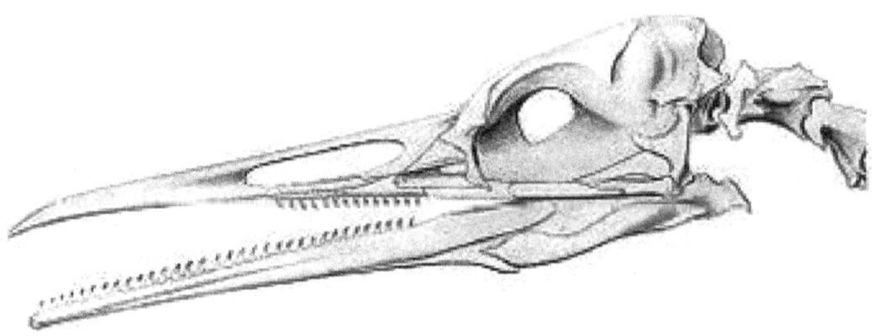

The Cretaceous seabird *Hesperornis*

While *Hesperornis* is not thought to have left descendants, the earliest modern seabirds also occurred in the Cretaceous, with a species called *Tytthostonyx glauconiticus*, which seems allied to the Procellariiformes and Pelecaniformes. In the Paleogene the seas were dominated by early Procellariidae, giant penguins and two extinct families, the Pelagornithidae and the Plotopteridae (a group of large seabirds that looked like the penguins). Modern genera began their wide radiation in the Miocene, although the genus *Puffinus* (which includes today's Manx shearwater and sooty shearwater) might date back to the Oligocene. The highest diversity of seabirds apparently existed during the Late Miocene and the Pliocene. At the end of the latter, the oceanic food web had undergone a period of upheaval due to extinction of considerable numbers of marine species; subsequently, the spread of marine mammals seems to have prevented seabirds from reaching their erstwhile diversity.

Characteristics

Adaptations to Life at Sea

Seabirds have made numerous adaptations to living on and feeding in the sea. Wing morphology has been shaped by the niche an individual species or family has evolved, so that looking at a wing›s shape and loading can tell a scientist about its life feeding behaviour. Longer wings and low wing loading are typical of more pelagic species, while diving species have shorter wings. Species such as the wandering albatross, which forage over huge areas of sea, have a reduced capacity for powered flight and are dependent on a type of gliding called dynamic soaring (where the wind deflected by waves provides lift) as well as slope soaring. Seabirds also almost always have webbed feet, to aid movement on the surface as well as assisting diving in some species. The Procellariiformes are unusual among birds in having a strong sense of smell, which is used to find widely distributed food in a vast ocean, and possibly to locate their colonies.

Salt glands are used by seabirds to deal with the salt they ingest by drinking and feeding (particularly on crustaceans), and to help them osmoregulate. The excretions from these glands (which are positioned in the head of the birds, emerging from the nasal cavity) are almost pure sodium chloride.

Cormorants, like this double-crested cormorant, have plumage that is partly wettable. This functional adaptation balances the competing requirement for thermoregulation against that of the need to reduce buoyancy.

With the exception of the cormorants and some terns, and in common with most other birds, all seabirds have waterproof plumage. However, compared to land birds, they have far more feathers protecting their bodies. This dense plumage is better able to protect the bird from getting wet, and cold is kept out by a dense layer of down feathers. The cormorants possess a layer of unique feathers that retain a smaller layer of air (compared to other diving birds) but otherwise soak up water. This allows them to swim without fighting the buoyancy that retaining air in the feathers causes, yet retain enough air to prevent the bird losing excessive heat through contact with water.

The plumage of most seabirds is less colourful than that of land birds, restricted in the main to variations of black, white or grey. A few species sport colourful plumes (such as the tropicbirds and some penguins), but most of the colour in seabirds appears in the bills and legs. The plumage of seabirds is thought in many cases to be for camouflage, both defensive (the colour of US Navy battleships is the same as that of Antarctic prions, and in both cases it reduces visibility at sea) and aggressive (the white underside possessed by many seabirds helps hide them from prey below).

Diet and Feeding

Seabirds evolved to exploit different food resources in the world's seas and oceans, and to a great extent, their physiology and behaviour have been shaped by their diet. These evolutionary forces have often caused species in different families and even orders to evolve similar strategies and adaptations to the same problems, leading to remarkable convergent evolution, such as that between auks and penguins. There are four basic feeding strategies, or ecological guilds, for feeding at sea: surface feeding, pursuit diving, plunge diving and predation of higher vertebrates; within these guilds there are multiple variations on the theme.

Surface Feeding

Many seabirds feed on the ocean's surface, as the action of marine currents often concentrates food such as krill, forage fish, squid or other prey items within reach of a dipped head.

Wilson's storm petrels pattering on the water's surface

Surface feeding itself can be broken up into two different approaches, surface feeding while flying (for example as practiced by gadfly petrels, frigatebirds and storm petrels), and surface feeding while swimming (examples of which are practiced by fulmars, gulls, many of the shearwaters and gadfly petrels). Surface feeders in flight include some of the most acrobatic of seabirds, which either snatch morsels from the water (as do frigatebirds and some terns), or "walk", pattering and hovering on the water's surface, as some of the storm-petrels do.Many of these do not ever land in the water, and some, such as the frigatebirds, have difficulty getting airborne again should they do so.Another seabird family that does not land while feeding is the skimmer, which has a unique fishing method: flying along the surface with the lower mandible in the water—this shuts automatically when the bill touches something in the water. The skimmer›s bill reflects its unusual lifestyle, with the lower mandible uniquely being longer than the upper one.

Surface feeders that swim often have unique bills as well, adapted for their specific prey. Prions have special bills with filters called lamellae to filter out plankton from mouthfuls of water, and many albatrosses and petrels have hooked bills to snatch fast-moving prey. Gulls have more generalised bills that reflect their more opportunistic lifestyle.

Pursuit Diving

The chinstrap penguin is a highly streamlined pursuit diver.

Pursuit diving exerts greater pressures (both evolutionary and physiological) on seabirds, but the reward is a greater area in which to feed than is available to surface feeders. Propulsion underwater can be provided by wings (as used by penguins, auks, diving petrels and some other species of petrel) or feet (as used by cormorants, grebes, loons and several types of fish-eating ducks). Wing-propelled divers are generally faster than foot-propelled divers. In both cases, the use of wings or feet for diving has limited their utility in other situations: loons and grebes walk with extreme difficulty (if at all), penguins cannot fly, and auks have sacrificed flight efficiency in favour of underwater diving. For example, the razorbill (an Atlantic auk) requires 64% more energy to fly than a petrel of equivalent size. Many shearwaters are intermediate between the two, having longer wings than typical wing-propelled divers but heavier wing loadings than the other surface-feeding procellariids, leaving them capable of diving to considerable depths while still being efficient long-distance travellers. The deepest diving exhibited by shearwaters is found in the short-tailed shearwater, which has been recorded diving below 70 m. Some albatross species are also capable of limited diving, with light-mantled sooty albatrosses holding the record at 12 m. Of all the wing-propelled pursuit divers, the most efficient in the air are the albatrosses, and they are also the poorest divers. This is the dominant guild in polar and subpolar environments, as it is energetically inefficient in warmer waters. With their poor flying ability, many wing-propelled pursuit divers are more limited in their foraging range than other guilds, especially during the breeding season when hungry chicks need regular feeding.

Plunge Diving

Gannets, boobies, tropicbirds, some terns and brown pelicans all engage in plunge diving, taking fast moving prey by diving into the water from flight. Plunge diving allows birds to use the energy from the momentum of the dive to combat natural buoyancy (caused by air trapped in plumage), and thus uses less energy than the dedicated pursuit divers, allowing them to utilise more widely distributed food resources, for example, in impoverished tropical seas. In general, this is the most specialised method of hunting employed by seabirds; other non-specialists (such as gulls and skuas) may employ it but do so with less skill and from lower heights. In brown pelicans the skills of plunge diving take several years to fully develop—once mature, they can dive from 20 m (70 ft) above the water's surface, shifting the body before impact to avoid injury. It has been suggested that plunge divers are restricted in their hunting grounds to clear waters that afford a view of their prey from the air, and while they are the dominant guild in the tropics, the link between plunge diving and water clarity is inconclusive. Some plunge divers (as well as some surface feeders) are dependent on dolphins and tuna to push shoaling fish up towards the surface.

Kleptoparasitism, Scavenging, and Predation

This catch-all category refers to other seabird strategies that involve the next trophic level up. Kleptoparasites are seabirds that make a part of their living stealing food of other seabirds. Most famously, frigatebirds and skuas engage in this behaviour, although gulls, terns and other species will steal food opportunistically. The nocturnal nesting behaviour of some seabirds has been interpreted as arising due to pressure from this aerial piracy. Kleptoparasitism is not thought to play a significant part of the diet of any species, and is instead a supplement to food obtained by hunting. A study of great frigatebirds stealing from masked boobies estimated that the frigatebirds could at most obtain 40% of the food they needed, and on average obtained only 5%. Many species of gull will feed on seabird and sea mammal carrion when the opportunity arises, as will giant petrels. Some species of albatross also engage in scavenging: an analysis of regurgitated squid beaks has shown that many of the squid eaten are too large to have been caught alive, and include mid-water species likely to be beyond the reach of albatrosses. Some species will also feed on other seabirds; for example, gulls, skuas and pelicans will often take eggs, chicks and even small adult seabirds from nesting colonies, while the giant petrels can kill prey up to the size of small penguins and seal pups.

Life History

Seabirds' life histories are dramatically different from those of land birds. In general, they are K-selected, live much longer (anywhere between 20 and 60 years), delay breeding for longer (for up to 10 years), and invest more effort into fewer young. Most species will only have one clutch a year, unless they lose the first (with a few exceptions, like the Cassin`s auklet), and many species (like the tubenoses and sulids), only one egg a year.

Northern gannet pair "billing" during courtship; like all seabirds except the phalaropes they maintain a pair bond throughout the breeding season.

Care of young is protracted, extending for as long as six months, among the longest for birds. For example, once common guillemot chicks fledge, they remain with the male parent for several months at sea. The frigatebirds have the longest period of parental care of any bird except a few raptors and the southern ground hornbill,with each chick fledging after four to six months and continued assistance after that for up to fourteen months. Due to the extended period of care, breeding occurs every two years rather than annually for some species. This life-history strategy has probably evolved both in response to the challenges of living at sea (collecting widely scattered prey items), the frequency of breeding failures due to unfavourable marine conditions, and the relative lack of predation compared to that of land-living birds.

Because of the greater investment in raising the young and because foraging for food may occur far from the nest site, in all seabird species except the phalaropes, both parents participate in caring for the young, and pairs are typically at least seasonally monogamous. Many species, such as gulls, auks and penguins, retain the same mate for several seasons, and many petrel species mate for life. Albatrosses and procellariids, which mate for life, take many years to form a pair bond before they breed, and the albatrosses have an elaborate breeding dance that is part of pair-bond formation.

Breeding and Colonies

Common murres breed on densely packed colonies on offshore rocks, islands and cliffs.

Ninety-five percent of seabirds are colonial, and seabird colonies are among the largest bird colonies in the world, providing one of Earth`s great wildlife spectacles. Colonies of over a million birds have been recorded, both in the tropics (such as Kiritimati in the Pacific) and in the polar latitudes (as in Antarctica). Seabird colonies occur exclusively for the purpose of breeding; non-breeding birds will only collect together outside the breeding season in areas where prey species are densely aggregated.

Seabird colonies are highly variable. Individual nesting sites can be widely spaced, as in an albatross colony, or densely packed as with a murre colony. In most seabird colonies, several different species will nest on the same colony, often exhibiting some niche separation. Seabirds can nest in trees (if any are available), on the ground (with or without nests), on cliffs, in burrows under the ground and in rocky crevices. Competition can be strong both within species and between species, with aggressive species such as sooty terns pushing less dominant species out of the most desirable nesting spaces. The tropical Bonin petrel nests during the winter to avoid competition with the more aggressive wedge-tailed shearwater. When the seasons overlap, the wedge-tailed shearwaters will kill young Bonin petrels in order to use their burrows.

Many seabirds show remarkable site fidelity, returning to the same burrow, nest or site for many years, and they will defend that site from rivals with great vigour. This increases breeding success, provides a place for returning mates to reunite, and reduces the costs of prospecting for a new site. Young adults breeding for the first time usually return to their natal colony, and often nest close to where they hatched. This tendency, known as philopatry, is so strong that a study of Laysan albatrosses found that the average distance between hatching site and the site where a bird established its own territory was 22 m; another study, this time on Cory›s shearwaters nesting near Corsica, found that of nine out of 61 male chicks that returned to breed at their natal colony bred in the burrow they were raised in, and two actually bred with their own mother.

Colonies are usually situated on islands, cliffs or headlands, which land mammals have difficulty accessing. This is thought to provide protection to seabirds, which are often very clumsy on land. Coloniality often arises in types of bird that do not defend feeding territories (such as swifts, which have a very variable prey source); this may be a reason why it arises more frequently in seabirds. There are other possible advantages: colonies may act as information centres, where seabirds returning to the sea to forage can find out where prey is by studying returning individuals of the same species. There are disadvantages to colonial life, particularly the spread of disease. Colonies also attract the attention of predators, principally other birds, and many species attend their colonies nocturnally to avoid predation.

Migration

Like many birds, seabirds often migrate after the breeding season. Of these, the trip taken by the Arctic tern is the farthest of any bird, crossing the equator in order to spend the Austral summer in Antarctica. Other species also undertake trans-equatorial trips, both from the north to the south, and from south to north. The population of elegant terns, which nest off Baja California, splits after the breeding season with some birds travelling north to the Central Coast of California and some travelling as far south as Peru and Chile to feed in the Humboldt Current. The sooty shearwater undertakes an annual migration cycle that rivals that of the Arctic tern; birds that nest in New Zealand and Chile and spend the northern summer feeding in the North Pacific off Japan, Alaska and California, an annual round trip of 40,000 statute miles (64,000 km).

Pelican flock flying over Havana Bay area. These birds come to Cuba every year from North America in the northern hemisphere winter season.

Arctic terns breed in the arctic and subarctic and winter in Antarctica.

Other species also migrate shorter distances away from the breeding sites, their distribution at sea determined by the availability of food. If oceanic conditions are unsuitable, seabirds will emigrate to more productive areas, sometimes permanently if the bird is young. After fledging, juvenile birds often disperse further than adults, and to different areas, so are commonly sighted far from a species› normal range. Some species, such as the auks, do not have a concerted migration effort, but drift southwards as the winter approaches. Other species, such as some of the storm petrels, diving petrels and cormorants, never disperse at all, staying near their breeding colonies year round.

Away from the sea

While the definition of seabirds suggests that the birds in question spend their lives on the ocean, many seabird families have many species that spend some or even most of their lives inland away from the sea. Most strikingly, many species breed tens, hundreds or even thousands of miles inland. Some of these species still return to the ocean to feed; for example, the snow petrel, the nests of which have been found 480 kilometres (300 mi) inland on the Antarctic mainland, are unlikely to find anything to eat around their breeding sites. The marbled murrelet nests inland in old growth forest, seeking huge conifers with large branches to nest on. Other species, such as the California gull, nest and feed inland on lakes, and then move to the coasts in the winter. Some cormorant, pelican, gull and tern species have individuals that never visit the sea at all, spending their lives on lakes, rivers, swamps and, in the case of some of the gulls, cities and agricultural land. In these cases it is thought that these terrestrial or freshwater birds evolved from marine ancestors. Some seabirds, principally those that nest in tundra, as skuas and phalaropes do, will migrate over land as well.

The more marine species, such as petrels, auks and gannets, are more restricted in their habits, but are occasionally seen inland as vagrants. This most commonly happens to young inexperienced birds, but can happen in great numbers to exhausted adults after large storms, an event known as a *wreck*, where they provide prized sightings for birders.

Relationship with Humans

Seabirds and fisheries

Seabirds have had a long association with both fisheries and sailors, and both have drawn benefits and disadvantages from the relationship.

Fishermen have traditionally used seabirds as indicators of both fish shoals, underwater banks that might indicate fish stocks, and of potential landfall. In fact, the known association of seabirds with land was instrumental in allowing the Polynesians to locate tiny landmasses in the Pacific. Seabirds have provided food for fishermen away from home, as well as bait. Famously, tethered cormorants have been used to catch fish directly. Indirectly, fisheries have also benefited from guano from colonies of seabirds acting as fertilizer for the surrounding seas. Negative effects on fisheries are mostly restricted to raiding by birds on aquaculture, although long-lining fisheries also have to deal with bait stealing. There have been claims of prey depletion by seabirds of fishery stocks, and while there is some evidence of this, the effects of seabirds are considered smaller than that of marine mammals and predatory fish (like tuna).

Seabirds (mostly northern fulmars) flocking at a long-lining vessel

Some seabird species have benefited from fisheries, particularly from discarded fish and offal. These discards compose 30% of the food of seabirds in the North Sea, for example, and compose up to 70% of the total food of some seabird populations.This can have other impacts; for example, the spread of the northern fulmar through the United Kingdom is attributed in part to the availability of discards. Discards generally benefit surface feeders, such as gannets and petrels, to the detriment of pursuit divers like penguins.

Fisheries also have negative effects on seabirds, and these effects, particularly on the long-lived and slow-breeding albatrosses, are a source of increasing concern to conservationists. The bycatch of seabirds entangled in nets or hooked on fishing lines has had a big impact on seabird numbers; for example, an estimated 100,000 albatrosses are hooked and drown each year on tuna lines set out by long-line fisheries. Overall, many hundreds of thousands of birds are trapped and killed each year, a source of concern for some of the rarest species (for example, only about 2,000 short-tailed albatrosses are known to still exist). Seabirds are also thought to suffer when overfishing occurs.

Exploitation

The hunting of seabirds and the collecting of seabird eggs have contributed to the declines of many species, and the extinction of several, including the great auk and the spectacled cormorant. Seabirds have been hunted for food by coastal peoples throughout history—one of the earliest instances known is in southern Chile, where archaeological excavations in middens has shown hunting of albatrosses, cormorants and shearwaters from 5000 BP. This pressure has led to some species becoming extinct in many places; in particular, at least 20 species of an original 29 no longer breed on Easter Island. In the 19th century, the hunting of seabirds for fat deposits and feathers for the millinery trade reached industrial levels. Muttonbirding (harvesting shearwater chicks) developed as important industries in both New Zealand and Tasmania, and the name of one species, the providence petrel, is derived from its seemingly miraculous arrival on Norfolk Island where it provided a windfall for starving European settlers. In the Falkland Islands, hundreds of thousands of penguins were harvested for their oil each year. Seabird eggs have also long been an important source of food for sailors undertaking long sea voyages, as well as being taken when settlements grow in areas near a colony. Eggers from San Francisco took almost half a million eggs a year from the Farallon Islands in the mid-19th century, a period in the islands' history from which the seabird species are still recovering.

Both hunting and egging continue today, although not at the levels that occurred in the past, and generally in a more controlled manner. For example, the Māori of Stewart Island/Rakiura continue to harvest the chicks of the sooty shearwater as they have done for centuries, using traditional methods (called *kaitiakitanga*) to manage the harvest, but now work with the University of Otago in studying the populations. In Greenland, however, uncontrolled hunting is pushing many species into steep decline.

Other threats

Other human factors have led to declines and even extinctions in seabird populations, colonies and species. Of these, perhaps the most serious are introduced species. Seabirds, breeding predominantly on small isolated islands, have lost many predator defence behaviours. Feral cats are capable of taking seabirds as large as albatrosses, and many introduced rodents, such as the Pacific rat, can take eggs hidden in burrows. Introduced goats, cattle, rabbits and other herbivores can lead to problems, particularly when species need vegetation to protect or shade their young. Disturbance of breeding colonies by humans is often a problem as well—visitors, even well-meaning tourists, can flush brooding adults off a colony leaving chicks and eggs vulnerable to predators.

This crested auklet was oiled in Alaska during the M/V Selendang Ayu spill of 2004.

The build-up of toxins and pollutants in seabirds is also a concern. Seabirds, being apex predators, suffered from the ravages of DDT until it was banned; among other effects, DDT was implicated in embryo development problems and the skewed sex ratio of western gulls in southern California. Oil spills are also a threat to seabird species, as both a toxin and because the feathers of the birds become saturated by the oil, causing them to lose their waterproofing. Oil pollution threatens species with restricted ranges or already depressed populations.

Conservation

The threats faced by seabirds have not gone unnoticed by scientists or the conservation movement. As early as 1903, U.S. President Theodore Roosevelt was convinced of the need to declare Pelican Island in Florida a National Wildlife Refuge to protect the bird colonies (including the nesting brown pelicans), and in 1909 he protected the Farallon Islands. Today many important seabird colonies are given some measure of protection, from Heron Island in Australia to Triangle Island in British Columbia.

Island restoration techniques, pioneered by New Zealand, enable the removal of exotic invaders from increasingly large islands. Feral cats have been removed from Ascension Island, Arctic foxes from many islands in the Aleutian Islands, and rats from Campbell Island. The removal of these introduced species has led to increases in numbers of species under pressure and even the return of extirpated ones. After the removal of cats from Ascension Island, seabirds began to nest there again for the first time in over a hundred years.

Seabird mortality caused by long-line fisheries can be greatly reduced by techniques such as setting long-line bait at night, dying the bait blue, setting the bait underwater, increasing the amount of weight on lines and by using bird scarers, and their deployment is increasingly required by many national fishing fleets. The international ban on the use of drift nets has also helped reduce the mortality of seabirds and other marine wildlife.

One of the Millennium Projects in the UK was the Scottish Seabird Centre, near the important bird sanctuaries on Bass Rock, Fidra and the surrounding islands. The area is home to huge colonies of gannets, puffins, skuas and other seabirds. The centre allows visitors to watch live video from the islands as well as learn about the threats the birds face and how we can protect them, and has helped to significantly raise the profile of seabird conservation in the UK. Seabird tourism can provide income for coastal communities as well as raise the profile of seabird conservation. For example, the northern royal albatross colony at Taiaroa Head in New Zealand attracts 40,000 visitors a year.

The plight of albatross and large seabirds, as well as other marine creatures, being taken as bycatch by long-line fisheries, has been addressed by a large number of non-governmental organizations (including BirdLife International, the American Bird Conservancy and the Royal Society for the Protection of Birds). This led to the Agreement on the Conservation of Albatrosses and Petrels, a legally binding treaty designed to protect these threatened species, which has been ratified by eleven countries as of 2008 (namely Argentina, Australia, Chile, Ecuador, France, New Zealand, Norway, Peru, South Africa, Spain and the United Kingdom).

Role in Culture

Depiction of a pelican with chicks on a stained glass window, Saint Mark's Church, Gillingham, Kent.

Many seabirds are little studied and poorly known, due to living far out to sea and breeding in isolated colonies. However, some seabirds, particularly, the albatrosses and gulls, have broken into popular consciousness. The albatrosses have been described as "the most legendary of birds", and have a variety of myths and legends associated with them, and today it is widely considered unlucky to harm them, although the notion that sailors believed that is a myth that derives from Samuel Taylor Coleridge›s famous poem, The Rime of the Ancient Mariner, in which a sailor is punished for killing an albatross by having to wear its corpse around his neck.

Instead of the Cross the Albatross

About my neck was hung

Sailors did, however, consider it unlucky to touch a storm petrel, especially one that has landed on the ship.

Gulls are one of the most commonly seen seabirds, given their use of human-made habitats (such as cities and dumps) and their often fearless nature. They therefore also have made it into the popular consciousness – they have been used metaphorically, as in *Jonathan Livingston Seagull* by Richard Bach, or to denote a closeness to the sea, such as their use in *The Lord of the Rings* – both in the insignia of Gondor and therefore Númenor (used in the design of the films), and to call Legolas to (and across) the sea. Other species have also made an impact; pelicans have long been associated with mercy and altruism because of an early Western Christian myth that they split open their breast to feed their starving chicks.

Seabird Families

The following are the groups of birds normally classed as seabirds.

Sphenisciformes (Antarctic and southern waters; 16 species)

- Spheniscidae penguins

Procellariiformes (Tubenoses: pan-oceanic and pelagic; 93 species)

- Diomedeidae albatrosses
- Procellariidae fulmars, prions, shearwaters, gadfly and other petrels
- Pelacanoididae diving petrels
- Hydrobatidae storm petrels

Pelecaniformes (Worldwide; 57 species)

- Pelecanidae pelicans
- Sulidae gannets and boobies
- Phalacrocoracidae cormorants
- Fregatidae frigatebirds
- Phaethontidae tropicbirds

Charadriiformes (Worldwide; 305 species, but only the families listed are classed as seabirds.)

- Stercorariidae skuas
- Laridae gulls

- Sternidae terns
- Rhynchopidae skimmers
- Alcidae auks

For an alternative taxonomy of these groups, Sibley-Ahlquist taxonomy.

Mammals

Sea otter, classic keystone species which controls sea urchin numbers

There are about 130 living and recently extinct marine mammal species such as seals, whales, manatees, sea otters and polar bears. They do not represent a distinct taxon or systematic grouping, but are instead unified by their reliance on the marine environment for feeding. Both cetaceans and sirenians are fully aquatic and therefore are obligate water dwellers. Seals and sea-lions are semiaquatic; they spend the majority of their time in the water, but need to return to land for important activities such as mating, breeding and molting. In contrast, both otters and the polar bear are much less adapted to aquatic living. Their diet varies considerably as well; some may eat zooplankton, others may eat fish, squid, shellfish, sea-grass and a few may eat other mammals.

Endangered blue whale, largest animal ever

Bottlenose dolphin, highest encephalization of any animal after humans

Dugong grazing on seagrass

Walrus

Polar bear

REFERENCES

1. Drogin, Bob (2 August 2009). "Mapping an ocean of species". Los Angeles Times. Retrieved 18 August 2009.
2. Paul, Gregory S. (2010). The Evolution of Dinosaurs and their World. The Princeton Field Guide to Dinosaurs. Princeton: Princeton University Press. p. 19.
3. Bortolotti, Dan (2008). Wild Blue: A Natural History of the World›s Largest Animal. St. Martin›s Press.
4. Agre, P. (2006) "The aquaporin water channels".Proceedings of the American Thoracic Society, 3 (1): 5–13.
5. Collins J. C. (1991) The Matrix of Life: A View of Natural Molecules from the Perspective of Environmental WaterMolecular Presentations. ISBN 9780962971907.
6. 7,000 m Class Remotely Operated Vehicle KAIKO 7000. Japan Agency for Marine-Earth Science and Technology (JAMSTEC). Retrieved 7 June 2008.
7. Charette, Matthew A.; Smith, Walter H. F. (June 2010). The Volume of Earth›s Ocean. Oceanography. 23 (2): 112–14. doi:10.5670/oceanog.2010.51. Archived from the original (PDF) on 2 August 2013. Retrieved 6 June 2013.
8. sphere depth of the ocean – hydrology. Encyclopedia Britannica. Retrieved 12 April 2015.
9. Third rock from the Sun – restless Earth. NASA›s Cosmos. Retrieved 12 April 2015.
10. Perlman, Howard (17 March 2014). The World›s Water. USGS Water-Science School. Retrieved 12 April 2015.
11. Kennish, Michael J. (2001). Practical handbook of marine science. Marine science series (3rd ed.). CRC Press. p. 35. ISBN 0-8493-2391-6.
12. Mullen, Leslie (11 June 2002). Salt of the Early Earth. NASA Astrobiology Magazine. Archived from the original on 22 July 2007. Retrieved 14 March 2007.
13. Morris, Ron M. Oceanic Processes. NASA Astrobiology Magazine. Archived from the original on 15 April 2009. Retrieved 14 March 2007.
14. Scott, Michon (24 April 2006). Earth›s Big heat Bucket. NASA Earth Observatory. Retrieved 14 March 2007.
15. Sample, Sharron (21 June 2005). Sea Surface Temperature. NASA. Retrieved 21 April 2007.
16. Age of the Earth. United States Geological Survey. 9 July 2007. Retrieved 2015-05-31.
17. Dalrymple 2001, pp. 205–221
18. Manhesa, Gérard; Allègre, Claude J.; Dupréa, Bernard; Hamelin, Bruno (May 1980). Lead isotope study of basic-ultrabasic layered complexes: Speculations about the age of the earth and primitive mantle characteristics. Earth and Planetary Science Letters. Amsterdam, the Netherlands: Elsevier. 47 (3): 370–

382. Bibcode:1980E&PSL..47..370M. doi:10.1016/0012-821X(80)90024-2. ISSN 0012-821X.

19. Schopf, J. William; Kudryavtsev, Anatoliy B.; Czaja, Andrew D.; Tripathi, Abhishek B. (5 October 2007). Evidence of Archean life: Stromatolites and microfossils. Precambrian Research. Amsterdam, the Netherlands: Elsevier. 158 (3–4): 141–155. doi:10.1016/j.precamres.2007.04.009. ISSN 0301-9268.

20. Raven & Johnson 2002, p. 68

21. Borenstein, Seth (13 November 2013). Oldest fossil found: Meet your microbial mom. Excite. Yonkers, N.Y.: Mindspark Interactive Network. Associated Press. Retrieved 2015-05-31.

22. Pearlman, Jonathan (13 November 2013). ›Oldest signs of life on Earth found›. The Daily Telegraph. London: Telegraph Media Group. Retrieved 2014-12-15.

23. Noffke, Nora; Christian, Daniel; Wacey, David; Hazen, Robert M. (16 November 2013). Microbially Induced Sedimentary Structures Recording an Ancient Ecosystem in the ca. 3.48 Billion-Year-Old Dresser Formation, Pilbara, Western Australia. Astrobiology. New Rochelle, N.Y.: Mary Ann Liebert, Inc. 13 (12): 1103–1124. Bibcode:2013AsBio..13.1103N. doi:10.1089/ast.2013.1030. ISSN 1531-1074. PMC 3870916. PMID 24205812.

24. Ohtomo, Yoko; Kakegawa, Takeshi; Ishida, Akizumi; et al. (January 2014). Evidence for biogenic graphite in early Archaean Isua metasedimentary rocks. Nature Geoscience. London: Nature Publishing Group. 7 (1): 25–28. Bibcode:2014NatGe...7...25O. doi:10.1038/ngeo2025. ISSN 1752-0894.

25. Borenstein, Seth (19 October 2015). Hints of life on what was thought to be desolate early Earth. Excite. Yonkers, N.Y.: Mindspark Interactive Network. Associated Press. Retrieved 2015-10-20.

26. Bell, Elizabeth A.; Boehnike, Patrick; Harrison, T. Mark; et al. (24 November 2015). Potentially biogenic carbon preserved in a 4.1 billion-year-old zircon (PDF). Proceedings of the National Academy of Sciences of the United States of America. Washington, D.C.: National Academy of Sciences. 112 (47): 14518–14521. doi:10.1073/pnas.1517557112. ISSN 0027-8424. PMC 4664351. PMID 26483481. Retrieved 2015-12-30.

27. Penny, David; Poole, Anthony (December 1999). The nature of the last universal common ancestor. Current Opinion in Genetics & Development. Amsterdam, the Netherlands: Elsevier. 9 (6): 672–677. doi:10.1016/S0959-437X(99)00020-9. ISSN 0959-437X. PMID 10607605.

28. Theobald, Douglas L. (13 May 2010). A formal test of the theory of universal common ancestry. Nature. London: Nature Publishing Group. 465 (7295): 219–222. Bibcode:2010Natur.465..219T. doi:10.1038/nature09014. ISSN 0028-0836. PMID 20463738.

29. Doolittle, W. Ford (February 2000). Uprooting the Tree of Life (PDF). Scientific American. Stuttgart: Georg von Holtzbrinck Publishing Group. 282 (2): 90–95. doi:10.1038/scientificamerican0200-90. ISSN 0036-8733. PMID 10710791. Archived from the original (PDF) on 2006-09-07. Retrieved 2015-04-05.

30. Peretó, Juli (March 2005). Controversies on the origin of life (PDF). International Microbiology. Barcelona: Spanish Society for Microbiology. 8 (1): 23–31. ISSN 1139-6709. PMID 15906258.

31. Joyce, Gerald F. (11 July 2002). The antiquity of RNA-based evolution. Nature. London: Nature Publishing Group. 418 (6894): 214–221. Bibcode:2002Natur.418..214J. doi:10.1038/418214a. ISSN 0028-0836. PMID 12110897.

32. Trevors, Jack T.; Psenner, Roland (December 2001). From self-assembly of life to present-day bacteria: a possible role for nanocells. FEMS Microbiology Reviews. Amsterdam, the Netherlands: Elsevier on behalf of the Federation of European Microbiological Societies. 25 (5): 573–582. doi:10.1111/j.1574-6976.2001.tb00592.x. ISSN 1574-6976. PMID 11742692.

33. Bapteste, Eric; Walsh, David A. (June 2005). Does the ‹Ring of Life› ring true?. Trends in Microbiology. Cambridge, Mass.: Cell Press. 13 (6): 256–261. doi:10.1016/j.tim.2005.03.012. ISSN 0966-842X. PMID 15936656.

34. Darwin 1859, p. 1

35. Doolittle, W. Ford; Bapteste, Eric (13 February 2007). Pattern pluralism and the Tree of Life hypothesis. Proceedings of the National Academy of Sciences of the United States of America. Washington, D.C.: National Academy of Sciences. 104 (7): 2043–2049. Bibcode:2007PNAS..104.2043D. doi:10.1073/pnas.0610699104. ISSN 0027-8424. PMC 1892968. PMID 17261804.

36. Kunin, Victor; Goldovsky, Leon; Darzentas, Nikos; Ouzounis, Christos A. (July 2005). The net of life: Reconstructing the microbial phylogenetic network. Genome Research. Cold Spring Harbor, N.Y.: Cold Spring Harbor Laboratory Press. 15 (7): 954–959. doi:10.1101/gr.3666505. ISSN 1088-9051. PMC 1172039. PMID 15965028.

37. Jablonski, David (25 June 1999). The Future of the Fossil Record. Science. Washington, D.C.: American Association for the Advancement of Science. 284 (5423): 2114–2116. doi:10.1126/science.284.5423.2114. ISSN 0036-8075. PMID 10381868.

38. Ciccarelli, Francesca D.; Doerks, Tobias; von Mering, Christian; et al. (3 March 2006). Toward Automatic Reconstruction of a Highly Resolved Tree of Life. Science. Washington, D.C.: American Association for the Advancement of Science. 311 (5765): 1283–1287. Bibcode:2006Sci...311.1283C. doi:10.1126/science.1123061. ISSN 0036-8075. PMID 16513982.

39. Mason, Stephen F. (6 September 1984). Origins of biomolecular handedness. Nature. London: Nature Publishing Group. 311 (5981): 19–23. Bibcode:1984Natur.311...19M. doi:10.1038/311019a0. ISSN 0028-0836. PMID 6472461.

40. Wolf, Yuri I.; Rogozin, Igor B.; Grishin, Nick V.; Koonin, Eugene V. (1 September 2002). Genome trees and the tree of life. Trends in Genetics. Cambridge, Mass.: Cell Press. 18 (9): 472–479. doi:10.1016/S0168-9525(02)02744-0. ISSN 0168-9525. PMID 12175808.

41. Varki, Ajit; Altheide, Tasha K. (December 2005). Comparing the human and chimpanzee genomes: searching for needles in a haystack. Genome Research. Cold Spring Harbor, N.Y.: Cold Spring Harbor Laboratory Press. 15 (12): 1746–1758. doi:10.1101/gr.3737405. ISSN 1088-9051. PMID 16339373.

42. Cavalier-Smith, Thomas (29 June 2006). Cell evolution and Earth history: stasis and revolution. Philosophical Transactions of the Royal Society B: Biological Sciences. London: Royal Society. 361 (1470): 969–1006. doi:10.1098/rstb.2006.1842. ISSN 0962-8436. PMC 1578732. PMID 16754610.

43. Schopf, J. William (29 June 2006). Fossil evidence of Archaean life. Philosophical Transactions of the Royal Society B: Biological Sciences. London: Royal Society. 361 (1470): 869–885. doi:10.1098/rstb.2006.1834. ISSN 0962-8436. PMC 1578735. PMID 16754604.

44. Altermann, Wladyslaw; Kazmierczak, Józef (November 2003). Archean microfossils: a reappraisal of early life on Earth. Research in Microbiology. Amsterdam, the Netherlands: Elsevier for the Pasteur Institute. 154 (9): 611–617. doi:10.1016/j.resmic.2003.08.006. ISSN 0923-2508. PMID 14596897.

45. Schopf, J. William (19 July 1994). Disparate rates, differing fates: tempo and mode of evolution changed from the Precambrian to the Phanerozoic. Proceedings of the National Academy of Sciences of the United States of America. Washington, D.C.: National Academy of Sciences. 91 (15): 6735–6742. Bibcode:1994PNAS...91.6735S. doi:10.1073/pnas.91.15.6735. ISSN 0027-8424. PMC 44277. PMID 8041691.

46. Poole, Anthony M.; Penny, David (January 2007). Evaluating hypotheses for the origin of eukaryotes. BioEssays. Hoboken, N.J.: John Wiley & Sons. 29 (1): 74–84. doi:10.1002/bies.20516. ISSN 0265-9247. PMID 17187354.

47. Dyall, Sabrina D.; Brown, Mark T.; Johnson, Patricia J. (9 April 2004). Ancient Invasions: From Endosymbionts to Organelles. Science. Washington, D.C.: American Association for the Advancement of Science. 304 (5668): 253–257. Bibcode:2004Sci...304..253D. doi:10.1126/science.1094884. ISSN 0036-8075. PMID 15073369.

48. Martin, William (October 2005). The missing link between hydrogenosomes and mitochondria. Trends in Microbiology. Cambridge, Mass.: Cell Press. 13 (10): 457–459. doi:10.1016/j.tim.2005.08.005. ISSN 0966-842X. PMID 16109488.

49. Lang, B. Franz; Gray, Michael W.; Burger, Gertraud (December 1999). Mitochondrial genome evolution and the origin of eukaryotes. Annual Review of Genetics. Palo Alto, Calif.: Annual Reviews. 33: 351–397. doi:10.1146/annurev.genet.33.1.351. ISSN 0066-4197. PMID 10690412.

50. McFadden, Geoffrey Ian (1 December 1999). "Endosymbiosis and evolution of the plant cell". Current Opinion in Plant Biology. Amsterdam, the Netherlands: Elsevier. 2 (6): 513–519. doi:10.1016/S1369-5266(99)00025-4. ISSN 1369-5266. PMID 10607659.

51. DeLong, Edward F.; Pace, Norman R. (1 August 2001). Environmental Diversity of Bacteria and Archaea. Systematic Biology. Oxford: Oxford University

Press on behalf of the Society of Systematic Biologists. 50 (4): 470–478. doi:10.1080/106351501750435040. ISSN 1063-5157. PMID 12116647.

52. Kaiser, Dale (December 2001). Building a multicellular organism. Annual Review of Genetics. Palo Alto, Calif.: Annual Reviews. 35: 103–123. doi:10.1146/annurev.genet.35.102401.090145. ISSN 0066-4197. PMID 11700279.

53. Zimmer, Carl (7 January 2016). Genetic Flip Helped Organisms Go From One Cell to Many. The New York Times. Retrieved 7 January 2016.

54. Valentine, James W.; Jablonski, David; Erwin, Douglas H. (1 March 1999). Fossils, molecules and embryos: new perspectives on the Cambrian explosion (PDF). Development. Cambridge: The Company of Biologists. 126 (5): 851–859. ISSN 0950-1991. PMID 9927587. Retrieved 2014-12-30.

55. Ohno, Susumu (January 1997). The reason for as well as the consequence of the Cambrian explosion in animal evolution. Journal of Molecular Evolution. New York: Springer-Verlag New York. 44 (Suppl. 1): S23–S27. doi:10.1007/PL00000055. ISSN 0022-2844. PMID 9071008.

56. Valentine, James W.; Jablonski, David (2003). Morphological and developmental macroevolution: a paleontological perspective. The International Journal of Developmental Biology. Bilbao, Spain: University of the Basque Country Press. 47 (7–8): 517–522. ISSN 0214-6282. PMID 14756327. Retrieved 2014-12-30.

57. Wellman, Charles H.; Osterloff, Peter L.; Mohiuddin, Uzma (2003). Fragments of the earliest land plants. Nature. 425 (6955): 282–285. Bibcode:2003Natur.425..282W. doi:10.1038/nature01884. PMID 13679913.

58. Barton, Nicholas (2007). Evolution. pp. 273–274. ISBN 9780199226320. Retrieved 30 September 2012.

59. Waters, Elizabeth R. (December 2003). Molecular adaptation and the origin of land plants. Molecular Phylogenetics and Evolution. Academic Press. 29 (3): 456–463. doi:10.1016/j.ympev.2003.07.018. ISSN 1055-7903. PMID 14615186.

60. Mayhew, Peter J. (August 2007). Why are there so many insect species? Perspectives from fossils and phylogenies. Biological Reviews. Cambridge: Cambridge University Press on behalf of the Cambridge Philosophical Society. 82 (3): 425–454. doi:10.1111/j.1469-185X.2007.00018.x. ISSN 1464-7931. PMID 17624962.

61. Carroll, Robert L. (May 2007). The Palaeozoic Ancestry of Salamanders, Frogs and Caecilians. Zoological Journal of the Linnean Society. Hoboken, N.J.: Wiley-Blackwell. 150 (Supplement s1): 1–140. doi:10.1111/j.1096-3642.2007.00246.x. ISSN 1096-3642.

62. Wible, John R.; Rougier, Guillermo W.; Novacek, Michael J.; Asher, Robert J. (21 June 2007). Cretaceous eutherians and Laurasian origin for placental mammals near the K/T boundary. Nature. London: Nature Publishing Group. 447 (7147): 1003–1006. Bibcode:2007Natur.447.1003W. doi:10.1038/nature05854. ISSN 0028-0836. PMID 17581585.

63. Witmer, Lawrence M. (28 July 2011). Palaeontology: An icon knocked from

its perch. Nature. London: Nature Publishing Group. 475 (7357): 458–459. doi:10.1038/475458a. ISSN 0028-0836. PMID 21796198.

64. Schloss, Patrick D.; Handelsman, Jo (December 2004). Status of the Microbial Census. Microbiology and Molecular Biology Reviews. Washington, D.C.: American Society for Microbiology. 68 (4): 686–691. doi:10.1128/MMBR.68.4.686-691.2004. ISSN 1092-2172. PMC 539005. PMID 15590780.

65. Miller & Spoolman 2012, p. 62

66. Mora, Camilo; Tittensor, Derek P.; Adl, Sina; et al. (23 August 2011). How Many Species Are There on Earth and in the Ocean?. PLoS Biology. San Francisco: Public Library of Science. 9 (8): e1001127. doi:10.1371/journal.pbio.1001127. ISSN 1545-7885. PMC 3160336. PMID 21886479.

67. Madigan M; Martinko J, eds. (2006). Brock Biology of Microorganisms (13th ed.). Pearson Education. p. 1096. ISBN 0-321-73551-X.

68. Rybicki EP (1990). The classification of organisms at the edge of life, or problems with virus systematics. South African Journal of Science. 86: 182–6. ISSN 0038-2353.

69. Lwoff A (1956). The concept of virus. Journal of General Microbiology. 17 (2): 239–53. doi:10.1099/00221287-17-2-239. PMID 13481308.

70. Wade, Nicholas (25 July 2016). Meet Luca, the Ancestor of All Living Things. New York Times. Retrieved 25 July 2016.

71. 2002 WHO mortality data Accessed 20 January 2007

72. Functions of global ocean microbiome key to understanding environmental changes. www.sciencedaily.com. University of Georgia. 10 December 2015. Retrieved 11 December 2015.

73. Suttle, C.A. (2005). Viruses in the Sea. Nature. 437 (9): 356–361. doi:10.1038/nature04160. PMID 16163346.

74. Suttle CA. Marine viruses—major players in the global ecosystem. Nature Reviews Microbiology. 2007;5(10):801–12. doi:10.1038/nrmicro1750. PMID 17853907.

75. Mann, NH (2005-05-17). The Third Age of Phage. PLoS Biology. United States: Public Library of Science. 3 (5): 753–755. doi:10.1371/journal.pbio.0030182. PMC 1110918. PMID 15884981.

76. Wommack, K. Eric; Russell T. Hill; Terri A. Muller; Rita R. Colwell (April 1996). Effects of sunlight on bacteriophage viability and structure. Applied and Environmental Microbiology. United States of America: American Society for Microbiology. 62 (4): 1336–1341. PMC 167899. PMID 8919794.

77. Krupovic M, Bamford DH (2007). Putative prophages related to lytic tailless marine dsDNA phage PM2 are widespread in the genomes of aquatic bacteria. BMC Genomics. 8: 236. doi:10.1186/1471-2164-8-236. PMC 1950889. PMID 17634101.

78. Xue H, Xu Y, Boucher Y, Polz MF (2012). High frequency of a novel filamentous

phage, VCY φ, within an environmental Vibrio cholerae population. Applied and Environmental Microbiology. 78 (1): 28–33. doi:10.1128/AEM.06297-11. PMC 3255608. PMID 22020507.

79. Roux S, Krupovic M, Poulet A, Debroas D, Enault F (2012). Evolution and diversity of the Microviridae viral family through a collection of 81 new complete genomes assembled from virome reads. PLOS ONE. 7 (7): e40418. doi:10.1371/journal.pone.0040418. PMC 3394797. PMID 22808158.

80. Kellogg, CA; JB Rose; SC Jiang; JM Thurmond & JH Paul (1995). Genetic diversity of related vibriophages isolated from marine environments around Florida and Hawaii, USA. Marine Ecology Progress Series. Germany: Inter-Research Science Center. 120 (1–3): 89–98. doi:10.3354/meps120089.

81. Suttle CA. Viruses in the sea. Nature. 2005;437(7057):356–61. doi:10.1038/nature04160. PMID 16163346. Bibcode:2005Natur.437..356S.

82. www.cdc.gov. Harmful Algal Blooms: Red Tide: Home[Retrieved 2014-12-19].

83. Staff (2014). The Biosphere. Aspen Global Change Institute. Retrieved 10 November 2014.

84. Takamia; et al. (1997). Microbial flora in the deepest sea mud of the Mariana Trench. FEMS Microbiology Letters. 152 (2): 279–285. doi:10.1111/j.1574-6968.1997.tb10440.x.

85. National Geographic, 2005

86. Choi, Charles Q. (17 March 2013). Microbes Thrive in Deepest Spot on Earth. LiveScience. Retrieved 17 March 2013.

87. Glud, Ronnie; Wenzhöfer, Frank; Middelboe, Mathias; Oguri, Kazumasa; Turnewitsch, Robert; Canfield, Donald E.; Kitazato, Hiroshi (17 March 2013). High rates of microbial carbon turnover in sediments in the deepest oceanic trench on Earth. Nature Geoscience. 6 (4): 284–288. Bibcode:2013NatGe...6..284G. doi:10.1038/ngeo1773. Retrieved 17 March 2013.

88. Oskin, Becky (14 March 2013). Intraterrestrials: Life Thrives in Ocean Floor. LiveScience. Retrieved 17 March 2013.

89. Morelle, Rebecca (15 December 2014). Microbes discovered by deepest marine drill analysed. BBC News. Retrieved 15 December 2014.

90. Takai K; Nakamura K; Toki T; Tsunogai U; et al. (2008). Cell proliferation at 122°C and isotopically heavy CH4 production by a hyperthermophilic methanogen under high-pressure cultivation. Proceedings of the National Academy of Sciences of the United States of America. 105 (31): 10949–54. doi:10.1073/pnas.0712334105. PMC 2490668. PMID 18664583.

91. Fox, Douglas (20 August 2014). Lakes under the ice: Antarctica›s secret garden. Nature. 512: 244–246. Bibcode:2014Natur.512..244F. doi:10.1038/512244a. Retrieved 21 August 2014.

92. Mack, Eric (20 August 2014). Life Confirmed Under Antarctic Ice; Is Space Next?. Forbes. Retrieved 21 August 2014.

93. Koonin EV, Senkevich TG, Dolja VV. The ancient Virus World and evolution of cells. Biology Direct. 2006;1:29. doi:10.1186/1745-6150-1-29. PMID 16984643.

94. Canchaya C, Fournous G, Chibani-Chennoufi S, Dillmann ML, Brüssow H. Phage as agents of lateral gene transfer. Current Opinion in Microbiology. 2003;6(4):417–24. doi:10.1016/S1369-5274(03)00086-9. PMID 12941415.

95. Rybicki, EP. The classification of organisms at the edge of life, or problems with virus systematics. South African Journal of Science. 1990;86:182–186.

96. Koonin, E. V.; Starokadomskyy, P. (7 March 2016). Are viruses alive? The replicator paradigm sheds decisive light on an old but misguided question.. Studies in History and Philosophy of Biological and Biomedical Sciences. doi:10.1016/j.shpsc.2016.02.016. PMID 26965225.

97. Iyer LM, Balaji S, Koonin EV, Aravind L. Evolutionary genomics of nucleo-cytoplasmic large DNA viruses. Virus Research. 2006;117(1):156–84. doi:10.1016/j.virusres.2006.01.009. PMID 16494962.

98. Sanjuán R, Nebot MR, Chirico N, Mansky LM, Belshaw R.Viral mutation rates. Journal of Virology. 2010;84(19):9733–48. doi:10.1128/JVI.00694-10. PMID 20660197.

99. In: Mahy WJ, Van Regenmortel MHV. Desk Encyclopedia of General Virology. Oxford: Academic Press; 2009. ISBN 0-12-375146-2. p. 28.

100. Are viruses alive? The replicator paradigm sheds decisive light on an old but misguided question. Studies in History and Philosophy of Biological and Biomedical Sciences. 2016. doi:10.1016/j.shpsc.2016.02.016. PMID 26965225.

101. Holmes EC. Viral evolution in the genomic age. PLoS Biology. 2007;5(10):e278. doi:10.1371/journal.pbio.0050278. PMID 17914905.

102. Wimmer E, Mueller S, Tumpey TM, Taubenberger JK.Synthetic viruses: a new opportunity to understand and prevent viral disease. Nature Biotechnology. 2009;27(12):1163 72. doi:10.1038/nbt.1593. PMID 20010599.

103. Wommack KE, Colwell RR. Virioplankton: viruses in aquatic ecosystems. Microbiology and Molecular Biology Reviews. 2000;64(1):69–114. doi:10.1128/MMBR.64.1.69-114.2000. PMID 10704475.

104. Bergh O, Børsheim KY, Bratbak G, Heldal M. High abundance of viruses found in aquatic environments. Nature. 1989;340(6233):467–8. doi:10.1038/340467a0. PMID 2755508. Bibcode:1989Natur.340..467B.

105. Lawrence CM, Menon S, Eilers BJ, et al.. Structural and functional studies of archaeal viruses. Journal of Biological Chemistry. 2009;284(19):12599–603. doi:10.1074/jbc.R800078200. PMID 19158076.

106. Prangishvili D, Forterre P, Garrett RA. Viruses of the Archaea: a unifying view. Nature Reviews Microbiology. 2006;4(11):837–48. doi:10.1038/nrmicro1527. PMID 17041631.

107. Prangishvili D, Garrett RA. Exceptionally diverse morphotypes and genomes of crenarchaeal hyperthermophilic viruses. Biochemical Society Transactions. 2004;32(Pt 2):204–8. doi:10.1042/BST0320204. PMID 15046572.

108. Forterre P, Philippe H. The last universal common ancestor (LUCA), simple or complex?. The Biological Bulletin. 1999;196(3):373–5; discussion 375–7. doi:10.2307/1542973. PMID 11536914.

109. Fredrickson JK, Zachara JM, Balkwill DL, Kennedy D, Li SM, Kostandarithes HM, Daly MJ, Romine MF, Brockman FJ (2004). Geomicrobiology of high-level nuclear waste-contaminated vadose sediments at the Hanford site, Washington state. Applied and Environmental Microbiology. 70 (7): 4230–41. doi:10.1128/AEM.70.7.4230-4241.2004. PMC 444790. PMID 15240306.

110. Woese CR, Kandler O, Wheelis ML (1990). Towards a natural system of organisms: proposal for the domains Archaea, Bacteria, and Eucarya. Proceedings of the National Academy of Sciences of the United States of America. 87 (12): 4576–9. Bibcode:1990PNAS...87.4576W. doi:10.1073/pnas.87.12.4576. PMC 54159. PMID 2112744.

111. Schopf JW (1994). Disparate rates, differing fates: tempo and mode of evolution changed from the Precambrian to the Phanerozoic. Proceedings of the National Academy of Sciences of the United States of America. 91 (15): 6735–42. Bibcode:1994PNAS...91.6735S. doi:10.1073/pnas.91.15.6735. PMC 44277. PMID 8041691.

112. DeLong EF, Pace NR (2001). Environmental diversity of bacteria and archaea. Systematic Biology. 50 (4): 470–8. doi:10.1080/106351501750435040. PMID 12116647.

113. Brown JR, Doolittle WF (1997). Archaea and the prokaryote-to-eukaryote transition. Microbiology and Molecular Biology Reviews. 61 (4): 456–502. PMC 232621. PMID 9409149.

114. Poole AM, Penny D (2007). Evaluating hypotheses for the origin of eukaryotes. BioEssays. 29 (1): 74–84. doi:10.1002/bies.20516. PMID 17187354.

115. Lang BF, Gray MW, Burger G (1999). Mitochondrial genome evolution and the origin of eukaryotes. Annual Review of Genetics. 33: 351–97. doi:10.1146/annurev.genet.33.1.351. PMID 10690412.

116. McFadden GI (1999). Endosymbiosis and evolution of the plant cell. Current Opinion in Plant Biology. 2 (6): 513–9. doi:10.1016/S1369-5266(99)00025-4. PMID 10607659.

117. Patrick J. Keeling (2004). Diversity and evolutionary history of plastids and their hosts. American Journal of Botany. 91 (10): 1481–1493. doi:10.3732/ajb.91.10.1481. PMID 21652304.

118. The largest Bacterium: Scientist discovers new bacterial life form off the African coast, Max Planck Institute for Marine Microbiology, 8 April 1999, archived from the original on 20 January 2010

119. List of Prokaryotic names with Standing in Nomenclature - Genus Thiomargarita

120. Bang C, Schmitz RA (2015). Archaea associated with human surfaces: not to be underestimated. FEMS Microbiology Reviews. 39 (5): 631–48. doi:10.1093/femsre/fuv010. PMID 25907112.

121. Archaea Online Etymology Dictionary. Retrieved 17 August 2016.

122. Pace NR (May 2006). Time for a change. Nature. 441 (7091): 289. Bibcode:2006Natur.441..289P. doi:10.1038/441289a. PMID 16710401.

123. Stoeckenius W (1 October 1981). Walsby›s square bacterium: fine structure of an orthogonal procaryote. Journal of Bacteriology. 148 (1): 352–60. PMC 216199. PMID 7287626.

124. Ohtaki A, Kida H, Miyata Y, Ide N, Yonezawa A, Arakawa T, Iizuka R, Noguchi K, Kita A, Odaka M, Miki K and Yohda M (2008) Structure and molecular dynamics simulation of archaeal prefoldin: the molecular mechanism for binding and recognition of nonnative substrate proteins J. Mol. Biol.376: 1130–1141.

125. Cavalier-Smith T (December 1993). Kingdom protozoa and its 18 phyla. Microbiological Reviews. 57 (4): 953–94. PMC 372943. PMID 8302218.

126. Corliss JO (1992). Should there be a separate code of nomenclature for the protists?. BioSystems. 28 (1–3): 1–14. doi:10.1016/0303-2647(92)90003-H. PMID 1292654.

127. Devreotes P (1989). Dictyostelium discoideum: a model system for cell-cell interactions in development. Science. 245 (4922): 1054–8. Bibcode:1989Sci...245.1054D. doi:10.1126/science.2672337. PMID 2672337.

128. Slapeta J, Moreira D, López-García P (2005). The extent of protist diversity: insights from molecular ecology of freshwater eukaryotes. Proceedings of the Royal Society B: Biological Sciences Journal. 272 (1576): 2073–81. doi:10.1098/rspb.2005.3195. PMC 1559898. PMID 16191619.

129. Moreira D, López-García P (2002). The molecular ecology of microbial eukaryotes unveils a hidden world (PDF). Trends in Microbiology. 10 (1): 31–8. doi:10.1016/S0966-842X(01)02257-0. PMID 11755083.

130. Hyde, K.D.; E.B.J. Jones; E. Leaño; S.B. Pointing; A.D. Poonyth; L.L.P. Vrijmoed (1998). Role of fungi in marine ecosystems. Biodiversity and Conservation. 7 (9): 1147–1161. doi:10.1023/A:1008823515157.

131. Kirk, P.M., Cannon, P.F., Minter, D.W. and Stalpers, J. Dictionary of the Fungi. Edn 10. CABI, 2008

132. Hyde, K.D.; E.B.J. Jones (1989). Spore attachment in marine fungi. Botanica Marina. 32: 205–218. doi:10.1515/botm.1989.32.3.205.

133. San-Martín, A.; S. Orejanera; C. Gallardo; M. Silva; J. Becerra; R. Reinoso; M.C. Chamy; K. Vergara; J. Rovirosa (2008). Steroids from the marine fungus Geotrichum sp. Journal of the Chilean Chemical Society. 53 (1): 1377–1378.

134. Jones, E.B.G., Hyde, K.D., & Pang, K.-L., eds. (2014). Freshwater fungi: and fungal-like organisms. Berlin/Boston: De Gruyter.

135. Jones, E.B.G.; Pang, K.-L., eds. (2012). Marine Fungi, and Fungal-like Organisms. Marine and Freshwater Botany. Berlin, Boston: De Gruyter (published August 2012). ISBN 978-3-11-026406-7. Retrieved 3 September 2015.

136. Distribution and Diversity of Planktonic Fungi in the West Pacific Warm Pool. doi:10.1371/journal.pone.0101523.s001.

137. Wang, G.; Wang, X.; Liu, X.; Li, Q. (2012). Diversity and biogeochemical function of planktonic fungi in the ocean. In Raghukumar, C. Biology of marine fungi. Berlin, Heidelberg: Springer-Verlag. pp. 71–88. doi:10.1007/978-3-642-23342-5. ISBN 978-3-642-23341-8. Retrieved 3 September 2015.

138. Damare, Samir; Raghukumar, Chandralata (2007-11-11). Fungi and Macroaggregation in Deep-Sea Sediments. Microbial Ecology. 56 (1): 168–177. doi:10.1007/s00248-007-9334-y. ISSN 0095-3628.

139. Kubanek, Julia; Jensen, Paul R.; Keifer, Paul A.; Sullards, M. Cameron; Collins, Dwight O.; Fenical, William (2003-06-10). Seaweed resistance to microbial attack: A targeted chemical defense against marine fungi. Proceedings of the National Academy of Sciences. 100 (12): 6916–6921. doi:10.1073/pnas.1131855100. ISSN 0027-8424. PMC 165804. PMID 12756301.

140. Gao, Zheng; Johnson, Zackary I.; Wang, Guangyi (2009-07-30). Molecular characterization of the spatial diversity and novel lineages of mycoplankton in Hawaiian coastal waters. The ISME Journal. 4 (1): 111–120. doi:10.1038/ismej.2009.87. ISSN 1751-7362.

141. Panzer, Katrin; Yilmaz, Pelin; Weiß, Michael; Reich, Lothar; Richter, Michael; Wiese, Jutta; Schmaljohann, Rolf; Labes, Antje; Imhoff, Johannes F. (2015-07-30). Identification of Habitat-Specific Biomes of Aquatic Fungal Communities Using a Comprehensive Nearly Full-Length 18S rRNA Dataset Enriched with Contextual Data. PLoS ONE. 10 (7): e0134377. doi:10.1371/journal.pone.0134377. PMC 4520555. PMID 26226014.

142. Gutierrez, Marcelo H; Pantoja, Silvio; Quinones, Renato a and Gonzalez, Rodrigo R. First record of flamentous fungi in the coastal upwelling ecosystem off central Chile. Gayana (Concepc.) [online]. 2010, vol.74, n.1, pp. 66-73. ISSN 0717-6538.

143. Sridhar, K.R. (2009). 10. Aquatic fungi – Are they planktonic?. Plankton Dynamics of Indian Waters. Jaipur, India: Pratiksha Publications. pp. 133–148.

144. Species of Higher Marine Fungi University of Mississippi. Retrieved 2012-02-05.

145. Freshwater and marine lichen-forming fungi Retrieved 2012-02-06.

146. Yuan X & Xiao S (2005). Lichen-Like Symbiosis 600 Million Years Ago. Science. 308: 1017–1020. doi:10.1126/science.1111347. PMID 15890881.

147. Jones, E. B. Gareth; Pang, Ka-Lai (2012-08-31). Marine Fungi: and Fungal-like Organisms. Walter de Gruyter. ISBN 9783110264067.

148. Gould 1989, pp. 124–136

149. Davidson, Michael W. (26 May 2005). Animal Cell Structure. Molecular Expressions. Tallahassee, Fla.: Florida State University. Retrieved 2008-09-03.

150. Jun-Yuan Chen; Oliveri, Paola; Feng Gao; et al. (1 August 2002). Precambrian Animal Life: Probable Developmental and Adult Cnidarian Forms from Southwest China (PDF). Developmental Biology. Amsterdam, the Netherlands: Elsevier. 248 (1): 182–196. doi:10.1006/dbio.2002.0714. ISSN 0012-1606. PMID 12142030. Retrieved 2015-02-04.

151. Grazhdankin, Dima (June 2004). Patterns of distribution in the Ediacaran biotas: facies versus biogeography and evolution. Paleobiology. Paleontological Society. 30 (2): 203–221. doi:10.1666/0094-8373(2004)030<0203:PODITE>2.0.CO;2. ISSN 0094-8373.

152. Seilacher, Adolf (August 1992). Vendobionta and Psammocorallia: lost constructions of Precambrian evolution. Journal of the Geological Society. London: Geological Society of London. 149 (4): 607–613. doi:10.1144/gsjgs.149.4.0607. ISSN 0016-7649. Retrieved 2015-02-04.

153. Martin, Mark W.; Grazhdankin, Dmitriy V.; Bowring, Samuel A.; et al. (5 May 2000). Age of Neoproterozoic Bilaterian Body and Trace Fossils, White Sea, Russia: Implications for Metazoan Evolution. Science. Washington, D.C.: American Association for the Advancement of Science. 288 (5467): 841–845. Bibcode:2000Sci...288..841M. doi:10.1126/science.288.5467.841. ISSN 0036-8075. PMID 10797002. Retrieved 2015-02-05.

154. Fedonkin, Mikhail A.; Waggoner, Benjamin M. (28 August 1997). The late Precambrian fossil Kimberella is a mollusc-like bilaterian organism. Nature. London: Nature Publishing Group. 388 (6645): 868–871. Bibcode:1997Natur.388..868F. doi:10.1038/42242. ISSN 0028-0836. Retrieved 2008-07-03.

155. Mooi, Rich; David, Bruno (December 1998). Evolution Within a Bizarre Phylum: Homologies of the First Echinoderms (PDF). American Zoologist. Society for Integrative and Comparative Biology. 38 (6): 965–974. doi:10.1093/icb/38.6.965. ISSN 1540-7063. Retrieved 2015-02-05.

156. McMenamin, Mark A. S. (September 2003). Spriggina is a trilobitoid ecdysozoan. Geoscience Horizons Seattle 2003. Abstracts with Programs. 35. Boulder, Colo.: Geological Society of America. p. 105. OCLC 249088612. Retrieved 2007-11-24. Paper No. 40-2 presented at the Geological Society of America's 2003 Seattle Annual Meeting (2–5 November 2003) on 2 November 2003, at the Washington State Convention Center.

157. Jih-Pai Lin; Gon, Samuel M., III; Gehling, James G.; et al. (2006). A Parvancorina-like arthropod from the Cambrian of South China. Historical Biology: An International Journal of Paleobiology. Taylor & Francis. 18 (1): 33–45. doi:10.1080/08912960500508689. ISSN 1029-2381.

158. Butterfield, Nicholas J. (December 2006). Hooking some stem-group ‹worms›: fossil lophotrochozoans in the Burgess Shale. BioEssays. Hoboken, N.J.: John Wiley & Sons. 28 (12): 1161–1166. doi:10.1002/bies.20507. ISSN 0265-9247. PMID 17120226.

159. Budd, Graham E. (February 2003). The Cambrian Fossil Record and the Origin of the Phyla (PDF). Integrative and Comparative Biology. Oxford: Oxford University Press for the Society for Integrative and Comparative Biology. 43 (1): 157–165. doi:10.1093/icb/43.1.157. ISSN 1557-7023. PMID 21680420. Retrieved 2015-02-06.

160. Budd, Graham E. (March 1996). The morphology of Opabinia regalis and the reconstruction of the arthropod stem-group. Lethaia. Hoboken, N.J.: Wiley-

Blackwell. 29 (1): 1–14. doi:10.1111/j.1502-3931.1996.tb01831.x. ISSN 0024-1164.

161. Marshall, Charles R. (May 2006). Explaining the Cambrian ‹Explosion› of Animals. Annual Review of Earth and Planetary Sciences. Palo Alto, Calif.: Annual Reviews. 34: 355–384. Bibcode:2006AREPS..34..355M. doi:10.1146/annurev.earth.33.031504.103001. ISSN 1545-4495. Retrieved 2015-02-06.
162. Valentine, James W. (2004). On the Origin of Phyla. Chicago: University Of Chicago Press. p. 7. ISBN 0-226-84548-6. Classifications of organisms in hierarchical systems were in use by the seventeenth and eighteenth centuries. Usually organisms were grouped according to their morphological similarities as perceived by those early workers, and those groups were then grouped according to their similarities, and so on, to form a hierarchy.
163. Valentine, James W (2004-06-18). On the Origin of Phyla. ISBN 9780226845487.
164. Erwin, Douglas; Valentine, James; Jablonski, David (1997). Recent fossil finds and new insights into animal development are providing fresh perspectives on the riddle of the explosion of animals during the Early Cambrian. American Scientist (March–April).
165. Budd, G.E.; Jensen, S. (May 2000). A critical reappraisal of the fossil record of the bilaterian phyla. Biological Reviews. 75 (2): 253–295. doi:10.1111/j.1469-185X.1999.tb00046.x. PMID 10881389.
166. Ponder, W.F.; Lindberg, D.R., eds. (2008). Phylogeny and Evolution of the Mollusca. Berkeley: University of California Press. p. 481. ISBN 978-0-520-25092-5.
167. Porifera (n.) Online Etymology Dictionary. Retrieved 18 August 2016.
168. Vacelet & Duport 2004, pp. 179–190.
169. Spongia Linnaeus, 1759. World Register of Marine Species. Retrieved 18 July 2012.
170. Rowland, S. M. & Stephens, T. (2001). Archaeocyatha: A history of phylogenetic interpretation. Journal of Paleontology. 75 (6): 1065–1078. doi:10.1666/0022-3360(2001)075<1065:AAHOPI>2.0.CO;2. JSTOR 1307076. Archived from the original on 6 December 2008.
171. Sperling, E. A.; Pisani, D.; Peterson, K. J. (1 January 2007). Poriferan paraphyly and its implications for Precambrian palaeobiology (PDF). 286 (1). Geological Society, London, Special Publications: 355–368. doi:10.1144/SP286.25. Retrieved 22 August 2012.
172. Zhang, Z.-Q. (2011). Animal biodiversity: An introduction to higher-level classification and taxonomic richness (PDF). Zootaxa. 3148: 7–12.
173. Nematostella vectensis v1.0. Genome Portal. University of California. Retrieved 2014-01-19.
174. Nematostella. Nematostella.org. Retrieved 2014-01-18.
175. Genikhovich, G. & U. Technau (2009). The Starlet sea anemone Nematostella vectensis: An anthozoan model organism for studies in comparative genomics and

functional evolutionary developmental biology. Cold Spring Harbor Protocols. 2009. doi:10.1101/pdb.emo129.

176. Where Does Our Head Come From? Brainless Sea Anemone Sheds New Light on the Evolutionary Origin of the Head. Science Daily. 2013-02-12. Retrieved 2014-01-18.

177. Sinigaglia, C.; et al. (2013). The bilaterian head patterning gene six3/6 controls aboral domain development in a cnidarian. PLoS Biology. 11 (2): e1001488. doi:10.1371/journal.pbio.1001488.

178. Bavestrello, Giorgio; Christian Sommer; Michele Sarà (1992). Bi-directional conversion in Turritopsis nutricula (Hydrozoa). Scientia Marina. 56 (2–3): 137–140.

179. Piraino, Stefano; F. Boero; B. Aeschbach; V. Schmid (1996). Reversing the life cycle: medusae transforming into polyps and cell transdifferentiation in Turritopsis nutricula (Cnidaria, Hydrozoa). Biological Bulletin. Biological Bulletin, vol. 190, no. 3. 190 (3): 302–312. doi:10.2307/1543022. JSTOR 1543022.

180. Aurelia aurita Encyclopedia of Life. Retrieved 18 August 2016.

181. Karleskint G, Richard Turner R and , James Small J (2012) Introduction to Marine Biology Cengage Learning, edition 4, page 445. ISBN 9781133364467.

182. Cornwall – Nature – Superstar Worm. BBC.

183. Mark Carwardine (1995) The Guinness Book of Animal Records. Guinness Publishing. p. 232.

184. The Persistent Parasites. Time. 1957-04-08.

185. Hargis, William J. (1985). Parasitology and pathology of marine organisms of the world ocean. National Oceanic and Atmospheric Administration.

186. http://plpnemweb.ucdavis.edu/nemaplex/General/animpara.htm

187. Lynne S. Garcia (1999). Classification of Human Parasites, Vectors, and Similar Organisms Clinical Infectious Diseases (PDF). 29: 734–6.

188. Hodda, M (2011). Phylum Nematoda Cobb, 1932. In: Zhang, Z.-Q. Animal biodiversity: An outline of higher-level classification and survey of taxonomic richness. Zootaxa. 3148: 63–95.

189. Zhang, Z (2013). Animal biodiversity: An update of classification and diversity in 2013. In: Zhang, Z.-Q. (Ed.) Animal Biodiversity: An Outline of Higher-level Classification and Survey of Taxonomic Richness (Addenda 2013). Zootaxa. 3703 (1): 5–11. doi:10.11646/zootaxa.3703.1.3.

190. Lambshead PJD (1993). Recent developments in marine benthic biodiversity research. Oceanis. 19 (6): 5–24.

191. Borgonie G, García-Moyano A, Litthauer D, Bert W, Bester A, van Heerden E, Möller C, Erasmus M, Onstott TC (June 2011). Nematoda from the terrestrial deep subsurface of South Africa. Nature. 474 (7349): 79–82. doi:10.1038/nature09974. PMID 21637257.

192. Danovaro R, Gambi C, Dell›Anno A, Corinaldesi C, Fraschetti S, Vanreusel A,

Vincx M, Gooday AJ (January 2008). Exponential decline of deep-sea ecosystem functioning linked to benthic biodiversity loss. Current Biology. 18 (1): 1–8. doi:10.1016/j.cub.2007.11.056. PMID 18164201. Lay summary – EurekAlert!.

193. Platt HM (1994). foreword. In Lorenzen S, Lorenzen SA. The phylogenetic systematics of freeliving nematodes. London: The Ray Society. ISBN 0-903874-22-9.
194. Sea Lily. Science Encyclopedia. Retrieved 5 September 2014.
195. Animal Diversity Web - Echinodermata. University of Michigan Museum of Zoology. Retrieved 26 August 2012.
196. Fox, Richard. Asterias forbesi. Invertebrate Anatomy OnLine. Lander University. Retrieved 14 June 2014.
197. Holsinger, K. (2005). Keystone species. Retrieved 10 May 2010, from http://darwin.eeb.uconn.edu/eeb310/lecture-notes/interactions/node2.html
198. Chapman, A.D. (2009). Numbers of Living Species in Australia and the World, 2nd edition. Australian Biological Resources Study, Canberra. Retrieved 12 January 2010. ISBN 978-0-642-56860-1 (printed); ISBN 978-0-642-56861-8 (online).
199. Hancock, Rebecca (2008). Recognising research on molluscs. Australian Museum. Retrieved 2009-03-09.
200. Barnes, R.S.K.; Calow, P.; Olive, P.J.W.; Golding, D.W. & Spicer, J.I. (2001). The Invertebrates, A Synthesis (3rd ed.). UK: Blackwell Science.
201. Welcome to CephBase. CephBase. Retrieved 29 January 2016.
202. Wilbur, Karl M.; Clarke, M.R.; Trueman, E.R., eds. (1985), The Mollusca, 12. Paleontology and neontology of Cephalopods, New York: Academic Press, ISBN 0-12-728702-7
203. Black, Richard (26 April 2008). Colossal squid out of the freezer. BBC News.
204. Ewen Callaway (2 June 2008). Simple-Minded Nautilus Shows Flash of Memory. New Scientist. Retrieved 7 March 2012.
205. Kathryn Phillips (15 June 2008). Living Fossil Memories (PDF). Journal of Experimental Biology. 211 (12): iii. doi:10.1242/jeb.020370.
206. Robyn Crook & Jennifer Basil (2008). A biphasic memory curve in the chambered nautilus, Nautilus pompilius L. (Cephalopoda: Nautiloidea) (PDF). Journal of Experimental Biology. 211 (12): 1992–1998. doi:10.1242/jeb.018531.
207. Ruppert, Edward E.; Fox, Richard, S.; Barnes, Robert D. (2004). Invertebrate Zoology (7th ed.). Cengage Learning. ISBN 81-315-0104-3.
208. Hayward, PJ (1996). Handbook of the Marine Fauna of North-West Europe. Oxford University Press. ISBN 0-19-854055-8.
209. Giribet, G.; Okusu, A, A.; Lindgren, A.R., A. R.; Huff, S.W., S. W.; Schrödl, M, M.; Nishiguchi, M.K., M. K. (May 2006). Evidence for a clade composed of molluscs with serially repeated structures: monoplacophorans are related to chitons (Free full text). Proceedings of the National Academy of Sciences of the

United States of America. 103 (20): 7723–7728. Bibcode:2006PNAS..103.7723G. doi:10.1073/pnas.0602578103. PMC 1472512. PMID 16675549.

210. Healy, J.M. (2001). The Mollusca. In Anderson, D.T. Invertebrate Zoology (2nd ed.). Oxford University Press. pp. 120–171. ISBN 0-19-551368-1.

211. Ruppert, Fox & Barnes (2004), pp. 518–522

212. http://firstlifeseries.com/learn-more/

213. Braddy, Simon J.; Poschmann, Markus; Tetlie, O. Erik (2007). Giant claw reveals the largest ever arthropod. Biology Letters. 4 (1): 106–109. doi:10.1098/rsbl.2007.0491. PMC 2412931. PMID 18029297.

214. Cressey, Daniel (21 November 2007). Giant sea scorpion discovered. Nature. doi:10.1038/news.2007.272. Retrieved 10 June 2013.

215. Patrick Kilday (28 September 2005). Mantis shrimp boasts most advanced eyes. The Daily Californian.

216. S. N. Patek & R. L. Caldwell (2005). Extreme impact and cavitation forces of a biological hammer: strike forces of the peacock mantis shrimp. Journal of Experimental Biology. 208 (19): 3655–3664. doi:10.1242/jeb.01831. PMID 16169943.

217. D. R. Currie & T. M. Ward (2009). South Australian Giant Crab (Pseudocarcinus gigas) Fishery (PDF). South Australian Research and Development Institute. Fishery Assessment Report for PIRSA. Retrieved 9 December 2013.

218. An ugly giant crab of Japan Popular science, 1920, 96 (6): 42.

219. Gewin, V. (2005). Functional genomics thickens the biological plot. PLoS Biology, 3 (6), e219. doi:10.1371/journal.pbio.0030219

220. Lancelet (amphioxus) genome and the origin of vertebrates Ars Technica, 19 June 2008.

221. Lemaire, P. (2011) "Evolutionary crossroads in developmental biology: the tunicates". Development, 138 (11): 2143–2152. doi:10.1242/dev.048975

222. Hopkins Gareth R.; Brodie Edmund D. Jr (2015). Occurrence of Amphibians in Saline Habitats: A Review and Evolutionary Perspective. Herpetological Monographs. 29 (1): 1–27. doi:10.1655/HERPMONOGRAPHS-D-14-00006.

223. Natchev, Nikolay; Tzankov, Nikolay; Geme, Richard (2011). Green frog invasion in the Black Sea: habitat ecology of thePelophylax esculentus complex (Anura, Amphibia) population in the region of Shablenska Tuzla lagoon in Bulgaria(PDF). Herpetology Notes. 4: 347–351.

224. DOCKER, Margaret F (2006). Bill Beamish›s Contributions to Lamprey Research and Recent Advances in the Field.. Guelph Ichthyology Reviews. 7. Retrieved 12 June 2014.

225. Hardisty, M. W.; Potter, I. C. (1971). Hardisty, M. W.; Potter, I. C., eds. The Biology of Lampreys (1st ed.). Academic Press. ISBN 9780123248015.

226. Gill, Howard S.; Renaud, Claude B.; Chapleau, François; Mayden, Richard L.; Potter, Ian C.; Douglas, M. E. Phylogeny of Living Parasitic Lampreys

(Petromyzontiformes) Based on Morphological Data. Copeia. 2003 (4): 687–703. doi:10.1643/IA02-085.1.

227. Myxini – University of California Museum of Paleontology
228. Nielsen, Julius; Hedeholm, Rasmus B.; Heinemeier, Jan; Bushnell, Peter G.; Christiansen, Jørgen S.; Olsen, Jesper; Ramsey, Christopher Bronk; Brill, Richard W.; Simon, Malene; Steffensen, Kirstine F.; Steffensen, John F. (2016). Eye lens radiocarbon reveals centuries of longevity in the Greenland shark (Somniosus microcephalus). Science. 353 (6300): 702–4. doi:10.1126/science.aaf1703. Lay summary – Sci News (12 August 2016).
229. Wroe, S.; Huber, D. R.; Lowry, M.; McHenry, C.; Moreno, K.; Clausen, P.; Ferrara, T. L.; Cunningham, E.; Dean, M. N.; Summers, A. P. (2008). Three-dimensional computer analysis of white shark jaw mechanics: how hard can a great white bite? (PDF). Journal of Zoology. 276 (4): 336–342. doi:10.1111/j.1469-7998.2008.00494.x.
230. Pimiento, Catalina; Dana J. Ehret; Bruce J. MacFadden; Gordon Hubbell (10 May 2010). Stepanova, Anna, ed. Ancient Nursery Area for the Extinct Giant Shark Megalodon from the Miocene of Panama. PLoS ONE. Panama: PLoS.org. 5 (5): e10552. Bibcode:2010PLoSO...510552P. doi:10.1371/journal.pone.0010552. PMC 2866656. PMID 20479893. Retrieved 12 May 2010.
231. Lambert, Olivier; Bianucci, Giovanni; Post, Klaas; de Muizon, Christian; Salas-Gismondi, Rodolfo; Urbina, Mario; Reumer, Jelle (1 July 2010). The giant bite of a new raptorial sperm whale from the Miocene epoch of Peru. Nature. Peru. 466 (7302): 105–108. Bibcode:2010Natur.466..105L. doi:10.1038/nature09067. PMID 20596020.
232. Black, Richard (11 June 2007). Sawfish protection acquires teeth. BBC News.
233. Marshall, A.; Bennett, M.B.; Kodja, G.; Hinojosa-Alvarez, S.; Galvan-Magana, F.; Harding, M.; Stevens, G. & Kashiwagi, T. (2011). Manta birostris. IUCN Red List of Threatened Species. Version 2012.1. International Union for Conservation of Nature. Retrieved 10 March 2013.
234. Modesto, S.P.; Anderson, J.S. (2004). The phylogenetic definition of Reptilia. Systematic Biology. 53 (5): 815–821. doi:10.1080/10635150490503026. PMID 15545258.
235. Gauthier, J.A.; Kluge, A.G.; Rowe, T. (1988). The early evolution of the Amniota. In Benton, M.J. The Phylogeny and Classification of the Tetrapods. 1. Oxford: Clarendon Press. pp. 103–155. ISBN 978-0-19-857705-8.
236. Laurin, M.; Reisz, R. R. (1995). A reevaluation of early amniote phylogeny (PDF). Zoological Journal of the Linnean Society. 113 (2): 165–223. doi:10.1111/j.1096-3642.1995.tb00932.x.
237. Modesto, S.P. (1999). Observations of the structure of the Early Permian reptile Stereosternum tumidum Cope. Palaeontologia Africana. 35: 7–19.
238. Rasmussen, Arne Redsted; Murphy, John C.; Ompi, Medy; Gibbons, J. Whitfield; Uetz, Peter (2011-11-08). Marine Reptiles. PLoS ONE. 6 (11): e27373. doi:10.1371/journal.pone.0027373. PMC 3210815. PMID 22087300.

239. Sander, P. Martin (2012). Reproduction in early amniotes. Science. 337 (6096): 806–808. doi:10.1126/science.1224301.

240. Stidworthy J. 1974. Snakes of the World. Grosset & Dunlap Inc. 160 pp. ISBN 0-448-11856-4.

241. Sea snakes at Food and Agriculture Organization of the United Nations. Accessed 7 August 2007.

242. Rasmussen, A.R., Murphy, J.C., Ompi, M., Gibbons, J.W. and Uetz, P. (2011) Marine reptiles. PLOS one, 6 (11): e27373.

243. Martill D.M. (1993). "Soupy Substrates: A Medium for the Exceptional Preservation of Ichthyosaurs of the Posidonia Shale (Lower Jurassic) of Germany". Kaupia - Darmstädter Beiträge zur Naturgeschichte, 2 : 77-97.

244. Gould,Stephen Jay (1993) "Bent Out of Shape" in Eight Little Piggies: Reflections in Natural History. Norton, 179–94. ISBN 9780393311396.

245. Sardine Run Shark Feeding Frenzy Phenomenon in Africa.

246. The Society for Marine Mammalogy›s Taxonomy Committee List of Species and subspecies. Society for Marine Mammalogy. October 2015. Retrieved 23 November 2015.

247. Blue whale. World Wide Fund For Nature. Retrieved 15 August 2016.

248. Marino, Lori (2004). Cetacean Brain Evolution: Multiplication Generates Complexity (PDF). International Society for Comparative Psychology. The International Society for Comparative Psychology (17): 1–16.

249. Lalli, C.; Parsons, T. (1993). Biological Oceanography: An Introduction. Butterworth-Heinemann. ISBN 0 7506 3384 0.

250. John Dolan (November 2012). Microzooplankton: the microscopic (micro) animals (zoo) of the plankton (PDF).

251. Harvey, Edmund Newton (1952). Bioluminescence. Academic Press.

252. Prentice, I.C. (2001). The carbon cycle and atmospheric carbon dioxide. Climate change 2001: the scientific basis: contribution of Working Group I to the Third Assessment Report of the Intergouvernmental Panel on Climate Change / Houghton, J.T. [edit.] Retrieved 31 May 2012.

253. DeLong, E.F.; Beja, O. (2010). The light-driven proton pump proteorhodopsin enhances bacterial survival during tough times. PLoS Biology. 8 (4). doi:10.1371/journal.pbio.1000359. e1000359.

254. Béja, O., Aravind, L., Koonin, E.V., Suzuki, M.T., Hadd, A., Nguyen, L.P., Jovanovich, S.B., Gates, C.M., Feldman, R.A., Spudich, J.L. and Spudich, E.N. (2000) Bacterial rhodopsin: evidence for a new type of phototrophy in the sea. Science, 289 (5486): 1902-1906. doi:10.1126/science.289.5486.1902

255. Interviews with Fellows: Ed Delong. American Academy of Microbiology. Retrieved 2 July 2016.

256. Bacteria with Batteries, Popular Science magazine, January 2001, Page 55.

257. McKinney 1997, p. 110

258. Stearns & Stearns 1999, p. x

259. Novacek, Michael J. (8 November 2014). Prehistory›s Brilliant Future. The New York Times. New York: The New York Times Company. ISSN 0362-4331. Retrieved 2014-12-25.

260. Nee, S. (2004). Extinction, slime, and bottoms. PLoS Biology. 2 (8): E272. doi:10.1371/journal.pbio.0020272. PMC 509315. PMID 15314670.

261. Ward, Peter D (2006). Impact from the Deep. Scientific American. 295 (4): 64–71. doi:10.1038/scientificamerican1006-64.

262. Desmond, Adrian (1975). The Discovery of Marine Transgressions and the Explanation of Fossils in Antiquity. American Journal of Science 275: 692–697.

263. McKirahan, Richard D. "Xenophanes of Colophon. Philosophy Before Socrates. Indianapolis: Hackett Publishing Company, 1994. 66. Print.

264. Grene MG, Grene M and Depew D (2004) The Philosophy of Biology: An Episodic History pages 1–34, Cambridge University Press .ISBN 9780521643801.

265. Boylan, Michael "Aristotle: Biology" Internet Encyclopedia of Philosophy. Retrieved 5 August 2016.

266. Lee, H. D. P. (1948) "Place-Names and the date of Aristotle's Biological Works". Classical Quarterly, 42 (3/4) :61–7

267. Singer, Charles. A short history of biology. Oxford 1931.

268. Emily Kearns, "Animals, knowledge about," in Oxford Classical Dictionary, 3rd ed., 1996, p. 92.

269. Carl T. Bergstrom; Lee Alan Dugatkin (2012). Evolution. Norton. p. 35. ISBN 978-0-393-92592-0.

270. Rhodes, Frank Harold Trevor (1 January 1974). Evolution. Golden Press. p. 7. ISBN 978-0-307-64360-5.

271. Lalli, Carol M., and Timothy R. Parsons. Introduction. Biological Oceanography: An Introduction. First Edition ed. Tarrytown, New York: Pergamon, 1993. 7-21. Print.

272. Menden-Deuer, Susanne. Course Info, OCG 561 Biological Oceanography.

273. Miller, Charles B.; Patricia A. Wheeler (2012). Biological Oceanography (Second ed. ed.). Chinchester, West Sussex: John Wiley & Sons.

CHAPTER 7

Marine Resources Conservation

Chapter Contents

- Overview
- Coral Reefs
- Human Impact
- Techniques
- Technology and Halfway Technology
- Laws and Treaties
- Organizations and Education
- Extinct and Endangered Species
- Overabundance
- List of Vulnerable Arthropods
- Marine Conservation Activism

Marine conservation focusses on limiting human-caused damage to marine ecosystems, and on restoring damaged marine ecosystems. Marine conservation also focusses on preserving vulnerable marine species.

Coral reefs have a great amount of biodiversity.

Marine resources conservation also known as marine conservation, is the protection and preservation of ecosystems in oceans and seas. Marine resources conservation focuses on limiting human-caused damage to marine ecosystems, restoring damaged marine ecosystems, and preserving vulnerable marine species.

OVERVIEW

Marine resources conservation is a response to biological issues such as extinction and marine habitats change. Marine conservation is the study of conserving physical and biological marine resources and ecosystem functions. This is a relatively new discipline. Marine conservationists rely on a combination of scientific principles derived from marine biology, oceanography, and fisheries science, as well as on human factors such as demand for marine resources and marine law, economics and policy in order to determine how to best protect and conserve marine species and ecosystems. Marine resources conservation can be seen as sub-discipline of conservation biology.

CORAL REEFS

Coral reefs are the epicenter of immense amounts of biodiversity, and are a key player in the survival of an entire ecosystem. They provide various marine animals with food, protection, and shelter which keep generations of species alive.Furthermore, coral reefs are an integral part of sustaining human life through serving as a food source (i.e., fish and mollusks) as well as a marine space for ecotourism which provides economic benefits. Also, humans are now conducting research regarding the use of corals as new potential sources for pharmaceuticals (i.e. steroids and anti-inflammatory drugs).

Unfortunately, because of the human impact on coral reefs, these ecosystems are becoming increasingly degraded and in need of conservation. The biggest threats include overfishing, destructive fishing practices, and sedimentation and pollution from land-based sources. This, in conjunction with increased carbon in oceans, coral bleaching, and diseases, means that there are no pristine reefs anywhere in the world. Up to 88% of coral reefs in Southeast Asia are now threatened, with 50% of those reefs at either «high» or «very high» risk of disappearing, which directly affects the biodiversity and survival of species dependent on coral.

This is especially harmful to island nations such as Samoa, Indonesia, and the Philippines, because many people there depend on the coral reef ecosystems to feed their families and to make a living. However, many fishermen are unable to catch as many fish as they used to, so they are increasingly using cyanide and dynamite in fishing, which further degrades the coral reef ecosystem. This perpetuation of bad habits simply leads to the further decline of coral reefs and therefore perpetuates the problem. One way of stopping this cycle is by educating the local community about why the conservation of marine spaces that include coral reefs is important. Once the local communities understand the personal stakes, then they will fight to preserve the reefs. Conserving coral reefs has many economic, social, and ecological benefits, not only for the people who live on these islands, but for people throughout the world.

HUMAN IMPACT

The deterioration of coral reefs is mainly linked to human activities – 88% of reefs are threatened through various reasons as listed above, including excessive amounts of CO_2 (carbon dioxide) emissions. Oceans absorb approximately 1/3 of the CO_2 produced by humans, which has detrimental effects on the marine environment. The increasing levels of CO_2 in oceans change the seawater chemistry by decreasing the pH. This process is also known as ocean acidification. Acidification negatively affects the carbonate buffering system and drops the carbonate saturation by 30%, which results in a decrease in reef calcification. Reductions in calcification have negative implications on calcifiers, such as corals and shellfish. Some examples include diminishing coral resilience from bleaching, decreasing organisms' ability to fight off predators, inhibiting their potential to compete for food, and altering behavior patterns. When the bottom of the food web declines tremendously due to acidification, the food web and the whole marine conservation effort is jeopardized. Although humans cause the greatest threat to the marine environment, they also have the ability to create effective management plans that will be the key to successful marine conservation. Although the most widely known conservation tool is the MPA, one of the best marine conservation tools simply stems from smarter individual choices made in efforts to reduce CO_2 emissions on a daily basis.

TECHNIQUES

Strategies and techniques for marine conservation tend to combine theoretical disciplines, such as population biology, with practical conservation strategies, such as setting up protected areas, as with marine protected areas (MPAs) or Voluntary Marine Conservation Areas. Other techniques include developing sustainable fisheries and restoring the populations of endangered species through artificial means.

Another focus of conservationists is on curtailing human activities that are detrimental to either marine ecosystems or species through policy, techniques such as fishing quotas, like those set up by the Northwest Atlantic Fisheries Organization, or laws such as those listed below. Recognizing the economics involved in human use of marine ecosystems is key, as is education of the public about conservation issues. This includes educating tourists that come to an area who might not be familiar with certain regulations regarding the marine habitat. One example of this is a project called Green Fins based in Southeast Asia that uses the scuba diving industry to educate the public. This project, implemented by UNEP, encourages scuba diving operators to educate their students about the importance of marine conservation and encourage them to dive in an environmentally friendly manner that does not damage coral reefs or associated marine ecosystems.

TECHNOLOGY AND HALFWAY TECHNOLOGY

Marine conservation technologies are used to protect endangered and threatened marine organisms and/or habitat. These technologies are innovative and revolutionary because they reduce bycatch, increase the survivorship and health of marine life and habitat, and benefit fishermen who depend on the resources for profit. Examples of technologies include marine protected areas (MPAs), turtle excluder devices (TEDs), autonomous recording unit, pop-up satellite archival tag, and radio-frequency identification (RFID). Commercial practicality plays an important role in the success of marine conservation because it is necessary to cater to the needs of fishermen while also protecting marine life.

Pop-up satellite archival tag (PSAT or PAT) plays a vital role in marine conservation by providing marine biologists with an opportunity to study animals in their natural environments. These are used to track movements of (usually large, migratory) marine animals. A PSAT is an archival tag (or data logger) that is equipped with a means to transmit the collected data via satellite. Though the data are physically stored on the tag, its major advantage is that it does not have to be physically retrieved like an archival tag for the data to be available, making it a viable independent tool for animal behavior studies. These tags have been used to track movements of ocean sunfish, marlin, blue sharks, bluefin tuna, swordfish and sea turtles. Location, depth, temperature, and body movement data are used to answer questions about migratory patterns, seasonal feeding movements, daily habits, and survival after catch and release.

Turtle excluder devices (TEDs) remove a major threat to turtles in their marine environment. Many sea turtles are accidentally captured, injured or killed by fishing. In response to this threat the National Oceanic and Atmospheric Administration (NOAA) worked with the shrimp trawling industry to create the TEDs. By working with the industry they insured the commercial viability of the devices. A TED is a series of bars that is placed at the top or bottom of a trawl net, fitting the bars into the «neck» of the shrimp trawl and acting as a filter to ensure that only small animals may pass through. The shrimp will be caught but larger animals such as marine turtles that become caught by the trawler will be rejected by the filter function of the bars.

Similarly, halfway technologies work to increase the population of marine organisms. However, they do so without behavioral changes, and address the symptoms but not the cause of the declines. Examples of halfway technologies include hatcheries and fish ladders.

LAWS AND TREATIES

International laws and treaties related to marine conservation include the 1966 Convention on Fishing and Conservation of Living Resources of the High Seas. United States laws related to marine conservation include the 1972 Marine Mammal Protection Act, as well as the 1972 Marine Protection, Research and Sanctuaries Act, which established the National Marine Sanctuaries program.

In 2010, the Scottish Parliament enacted new legislation for the protection of marine life with the Marine (Scotland) Act 2010. Its provisions include marine planning, marine licensing, marine conservation, seal conservation, and enforcement.

ORGANIZATIONS AND EDUCATION

There are marine conservation organizations throughout the world that focus on funding conservation efforts, educating the public and stakeholders, and lobbying for conservation law and policy. Examples of these include Oceana, the Marine Conservation Institute

(United States), Blue Frontier Campaign (United States), Sea Shepherd Conservation Society (international), Frontier (the Society for Environmental Exploration) (United Kingdom), Marine Conservation Society (United Kingdom), Community Centred Conservation (C3), the Reef-World Foundation (United Kingdom), Reef Watch (India), and Australian Marine Conservation Society. Zoox (United Kingdom) is an example of an organization that provides both marine conservation training and professional career development to volunteers who are also working on marine conservation projects such as Green Fins.

The shore of the Pacific Ocean in San Francisco, California

On a regional level, PERSGA, the Regional Organization for the Conservation of the Environment of the Red Sea and the Gulf of Aden, is a regional entity which serves as the secretariat for the Jeddah Convention-1982, one of the first regional marine agreements. PERSGA member states are Djibouti, Egypt, Jordan, Saudi Arabia, Somalia, Sudan, and Yemen.

EXTINCT AND ENDANGERED SPECIES

Marine Mammals

Baleen whales were predominantly hunted from 1600 through the mid-1900s, and were nearing extinction when a global ban on commercial whaling was put into effect in 1896 by the IWC (International Whaling Convention). The Atlantic gray whale, last sighted in 1740, is now extinct due to European and Native American Whaling. Since the 1960s the global population of monk seals has been rapidly declining. The Hawaiian and Mediterranean monk seals are considered to be one of the most endangered marine mammals on the planet, according to the NOAA. The last sighting of the Caribbean monk seal was in 1952, and it has now been confirmed extinct by the NOAA. The vaquita porpoise, discovered in 1958, has become the most endangered marine species. Over half the population has disappeared since 2012, leaving 100 left in 2014. The vaquita frequently drowns in fishing nets, which are used illegally in marine protected areas off the Gulf of Mexico.

Sea turtles

In 2004, the Marine Turtle Specialist Group (MTSG), from the International Union for Conservation of Nature (IUCN), ran an assessment which determined that green turtles were globally endangered. Population decline in ocean basins is indicated through data collected by the MTSG that analyzes abundance and historical information on the species. This data examined the global population of green turtles at 32 nesting sites, and determined that over the last 100–150 years there has been a 48-65 percent decrease in the number of mature nesting females. The Kemp›s ridley sea turtle population fell in 1947 when 33,000 nests, which accounted for 80 percent of the population, were collected and sold by villagers in Racho Nuevo, Mexico. In the early 1960s only 5,000 individuals were left, and between 1978 and 1991, 200 Kemp's ridley turtles nested annually. In 2015, the World Wildlife Fund and *National Geographic Magazine* named the Kemp's ridley the most endangered sea turtle in the world, with 1000 females nesting annually.

Fish

In 2014, the IUCN moved the Pacific bluefin tuna from "least concerned" to "vulnerable" on a scale that represents level of extinction risk. The Pacific bluefin tuna is targeted by the fishing industry mainly for its use in sushi. A stock assessment released in 2013 by the International Scientific Committee for Tuna and Tuna-Like Species in the North Pacific Ocean (ISC) shows that the Pacific bluefin tuna population dropped by 96 % in the Pacific Ocean. According to the ISC assessment, 90 % of the Pacific bluefin tuna caught are juveniles that have not reproduced.

Between 2011 and 2014, the European eel, Japanese eel, and American eel were put on the IUCN red list of endangered species. In 2015, the Environmental Agency concluded that the number of European eels has declined by 95 % since 1990. An Environmental Agency officer, Andy Don, who has been researching eels for the past 20 years, said, "There is no doubt that there is a crisis. People have been reporting catching 1 kg of glass eels this year when they would expect to catch 40 kg. We have got to do something."

Marine Plants

Johnson's seagrass, a food source for the endangered green sea turtle, is the scarcest species in its genus. It reproduces asexually, which limits its ability to populate and colonize habitats. Data on this species is limited, but it is known that since the 1970s there has been a 50 % decrease in abundance.

History of marine resources conservation

Modern marine resources conservation first became globally recognized in the 1970s after World War II in an era known as the "marine revolution". The United States federal legislation showed its support of marine resources conservation by institutionalizing protected areas and creating marine estuaries. In the mid-1970s the United States formed the International Union for Conservation of Nature (IUCN). Through this program, nations could communicate and make agreements about marine resources conservation. After the formation of the IUCN, new independent organizations known as non-governmental organizations started to appear. These organizations were self-governed and had individual goals for marine conservation. At the end of the 1970s, undersea explorations equipped with new technology such as computers were undertaken. During these explorations, fundamental principles of change were discovered in relation to marine ecosystems. Through this discovery, the interdependent nature of the ocean was revealed. This led to a change in the approach of marine conservation efforts, and a new emphasis was put on restoring systems within the environment, along with protecting biodiversity.

OVERABUNDANCE

Overabundance occurs when the population of a certain species cannot be controlled. The domination of one species can create an imbalance in an ecosystem, which can lead to the demise of other species and of the habitat. Overabundance occurs predominately in invasive species. Cargo ships introduce new species into different environments through releasing ballast water into an ecosystem. A tank of ballast water is estimated to contain around 3,000 non-native species.

The San Francisco Bay is one of the places in the world that is the most impacted by foreign and invasive species. According to the Baykeeper organization, 97 percent of the organisms in the San Francisco Bay have been compromised by the 240 invasive species that have been brought into the ecosystem. Invasive species in the bay such as the Asian clam (Corbicula fluminea) have changed the food web of the ecosystem by depleting populations of native species such as plankton. The Asian clam clogs pipes and obstructs the flow of water in electrical generating facilities. Their presence in the San Francisco Bay has cost the United States an estimated one billion dollars in damages.

LIST OF VULNERABLE ARTHROPODS

As of July 2016, the International Union for Conservation of Nature (IUCN) lists 1080 vulnerable arthropod species. 11% of all evaluated arthropod species are listed as vulnerable. The IUCN also lists 29 arthropod subspecies as vulnerable.

No subpopulations of arthropods have been evaluated by the IUCN.

For a species to be assessed as vulnerable to extinction the best available evidence must meet quantitative criteria set by the IUCN designed to reflect «a high risk of extinction in the wild». *Endangered* and *critically endangered* species also meet the quantitative criteria of *vulnerable* species, and are listed separately. See: List of endangered arthropods, List of critically endangered arthropods. Vulnerable, endangered and critically endangered species are collectively referred to as *threatened species* by the IUCN.

Additionally 2875 arthropod species (30% of those evaluated) are listed as *data deficient*, meaning there is insufficient information for a full assessment of conservation status. As these species typically have small distributions and/or populations, they are intrinsically likely to be threatened, according to the IUCN.While the category of *data deficient* indicates that no assessment of extinction risk has been made for the taxa, the IUCN notes that it may be appropriate to give them "the same degree of attention as threatened taxa, at least until their status can be assessed."

This is a complete list of vulnerable arthropod species and subspecies as evaluated by the IUCN.

- Fabaeformiscandona aemonae
- Leucocythere helenae
- Limnocythere porphyretica
- Newnhamia fuscata
- Newnhamia insolita
- Pseudocandona cavicola
- Pseudocandona pretneri
- Pseudocandona trigonelia
- Zonocypretta kalimna

Arachnids

There are 47 arachnid species assessed as vulnerable.

Spiders

- *Andasta benoiti*
- *Argyrodella pusillus*
- *Argyrodes cognatus*
- *Argyrodes fissifrontellus*
- *Baviola braueri*
- *Baviola vanmoli*
- *Brignolia trichinalis*
- *Carrhotus bellus*
- *Cenemus culiculus*
- *Cousinea keeleyi*
- Great raft spider *(Dolomedes plantarius)*
- *Goleba pallens*
- *Grammostola vachoni*
- *Hasarius rufociliatus*
- *Mesida thorelli*
- Dolloff cave spider *(Meta dolloff)*
- *Nesiergus insulanus*
- *Ouette ouette*
- Jean's jumping spider *(Paraheliophanus jeanae)*
- St Helenian jumping spider *(Paraheliophanus sanctaehelenae)*
- Under-equipped jumping spider *(Paraheliophanus subinstructus)*

- Glacier Bay wolf spider *(Pardosa diuturna)*
- *Phycosoma spundana*
- Mysore ornamental *(Poecilotheria striata)*
- *Rhomphaea barycephala*
- *Roche roche*
- *Sadies fulgida*
- *Sadies seychellensis*
- *Silhouettella curieusei*
- *Soeuria soeur*
- Lake Placid funnel wolf spider *(Sosippus placidus)*
- *Stoda libudum*
- Notable large burrowing spider *(Thrigmopoeus insignis)*
- *Troglohyphantes gracilis*
- *Troglohyphantes similis*
- *Troglohyphantes spinipes*
- Xerophaeus espoir

Other Arachnid Species

- *Bamazomus aviculus*
- Melones cave harvestman *(Banksula melones)*
- *Beierolpium benoiti*
- Seychelles small whip spider *(Charinus seychellarum)*
- Seychelles giant scorpion *(Chiromachus ochropus)*
- Empire cave pseudoscorpion *(Fissilicreagris imperialis)*
- *Ibalonius flavopictus*
- *Nesowithius seychellesensis*
- Indian Ocean whip spider *(Phrynichus scaber)*
- *Xenolpium insulare*

Branchiopoda

- *Alona hercegovinae*
- *Alona sketi*
- *Alona smirnovi*
- Vernal pool fairy shrimp *(Branchinecta lynchi)*
- *Branchinella apophysata*
- Branchinella basispina

- Branchinella denticulata
- Branchinella simplex
- Branchinella wellardi
- Chirocephalus croaticus
- Chirocephalus pelagonicus
- Chirocephalus reiseri
- Daphnia coronata
- Daphnia jollyi
- Daphnia nivalis
- Daphnia occidentalis
- Eoleptestheria spinosa
- Imnadia banatica
- Imnadia cristata
- Imnadia panonica
- Parartemia contracta
- Rhynchochydorus australiensis

Millipedes

- Solitary black millipede *(Doratogonus avius)*
- Bearded black millipede *(Doratogonus barbatus)*
- Herbert's black millipede *(Doratogonus herberti)*
- Hoffman's black millipede *(Doratogonus hoffmani)*
- Southern black millipede *(Doratogonus meridionalis)*
- Natal black millipede *(Doratogonus natalensis)*
- Precarious black millipede *(Doratogonus precarius)*
- *Eutrichodesmus griseus*
- *Hyperothrix orophura*
- *Plusioglyphiulus boutini*
- *Pterozonium braueri*
- *Rhinotus crassiceps*

Maxillopoda

Maxillopoda includes barnacles, copepods and a number of related animals. There are 71 species in the class Maxillopoda assessed as vulnerable.

Calanoida

There are 47 species in the order Calanoida assessed as vulnerable.

Diaptomids

- Aglaodiaptomus kingsburyae
- Aglaodiaptomus marshianus
- Allodiaptomus satanas
- Arctodiaptomus burduricus
- Arctodiaptomus euacanthus
- Arctodiaptomus kamtschaticus
- Arctodiaptomus michaeli
- Dussartius baeticus
- Eodiaptomus lumholtzi
- Eodiaptomus shihi
- Heliodiaptomus kolleruensis
- Heliodiaptomus pulcher
- Hesperodiaptomus augustaensis
- Hesperodiaptomus californiensis
- Lovenula excellens
- Lovenula simplex
- Mastigodiaptomus purpureus
- Metadiaptomus capensis
- Metadiaptomus purcelli
- Neodiaptomus intermedius
- Neodiaptomus laii
- Neodiaptomus lymphatus
- Neodiaptomus physalipus
- Notodiaptomus dubius
- Notodiaptomus maracaibensis
- Paradiaptomus natalensis
- Phyllodiaptomus wellekensae
- Skistodiaptomus carolinensis
- Stygodiaptomus kieferi
- Stygodiaptomus petkovskii
- Thermodiaptomus galeboides
- Tropodiaptomus burundensis

- Tropodiaptomus kilimensis
- Tropodiaptomus kissi
- Tropodiaptomus neumanni
- Tropodiaptomus simplex
- Tropodiaptomus stuhlmanni

Centropagids

- Boeckella bispinosa
- Boeckella calcaris
- Boeckella geniculata
- Boeckella nyoraensis
- Boeckella shieli
- Calamoecia australica
- Calamoecia elongata
- Calamoecia zeidleri
- Hemiboeckella powellensis

Temorids

- Epischura baikalensis

Cyclopoida

- Acanthocyclops hypogeus
- Mesocyclops insulensis
- Metacyclops gasparoi
- Metacyclops postojnae
- Tropocyclops federensis
- Tropocyclops nananae

Harpacticoida

There are 18 species in the order Harpacticoida assessed as vulnerable.

Darcythompsoniids

- Leptocaris stromatolicolus

Ameirids

- Nitocrella slovenica

- Nitocrella stochi

Canthocamptids

- Canthocamptus dedeckkeri
- Canthocamptus echinopyge
- Canthocamptus longipes
- Canthocamptus mammillifurca
- Canthocamptus sublaevis
- Canthocamptus tasmaniae
- Ceuthonectes rouchi
- Elaphoidella amabilis
- Elaphoidella franci
- Elaphoidella jeanneli
- Elaphoidella kieferi
- Fibulacamptus bisetosus
- Fibulacamptus gracilior
- Paramorariopsis anae
- Pseudomoraria triglavensis

Malacostracans

Malacostraca includes crabs, lobsters, crayfish, shrimp, krill, woodlice, and many others. There are 307 malacostracan species and 14 malacostracan subspecies assessed as vulnerable.

Isopods

Species

- Madison cave isopod *(Antrolana lira)*
- *Arubolana imula*
- Bat cave isopod *(Caecidotea macropoda)*
- Nickajack cave isopod *(Caecidotea nickajackensis)*
- *Calconiscellus gotscheensis*
- *Haplophthalmus abbreviatus*
- *Haplophthalmus rhinoceros*
- Rye Cove cave isopod *(Lirceus culveri)*
- *Metatrichoniscoides celticus*
- *Mexilana saluposi*

- *Mexistenasellus nulemex*
- Parzefall's stenasellid *(Mexistenasellus parzefalli)*
- Wilken's stenasellid *(Mexistenasellus wilkensi)*
- *Monolistra calopyge*
- *Monolistra schottlaenderi*
- *Moserius percoi*
- *Onchotelson brevicaudatus*
- *Onchotelson spatulatus*
- *Proasellus parvulus*
- *Proasellus slovenicus*
- *Sphaerolana karenae*
- *Sumatrillo sp. nov. ‹HC - blind›*
- *Uramphisopus pearsoni*

Subspecies

- Androniscus stygius dentatus
- Asellus aquaticus cavernicolus
- Proasellus pavani orientalis

Amphipods

There are 56 amphipod species and four amphipod subspecies assessed as vulnerable.

Hadziids

Subspecies

- Hadzia fragilis stochi

Paramelitids

- Paramelita flexa

Gammarids

- Bousfield's amphipod *(Gammarus bousfieldi)*
- Diminutive amphipod *(Gammarus hyalelloides)*
- Pecos amphipod *(Gammarus pecos)*

Crangonyctids

- Central Missouri cave amphipod *(Allocrangonyx hubrichti)*
- Oklahoma cave amphipod *(Allocrangonyx pellucidus)*
- Florida cave amphipod *(Crangonyx grandimanus)*
- Hobb's cave amphipod *(Crangonyx hobbsi)*
- Tidewater interstitial amphipod *(Stygobromus araeus)*
- Arizona cave amphipod *(Stygobromus arizonensis)*
- Balcones cave amphipod *(Stygobromus balconis)*
- Barr's cave amphipod *(Stygobromus barri)*
- Bifurcated cave amphipod *(Stygobromus bifurcatus)*
- Bowman's cave amphipod *(Stygobromus bowmani)*
- Clanton's cave amphipod *(Stygobromus clantoni)*
- Burnsville Cove cave amphipod *(Stygobromus conradi)*
- Cooper's cave amphipod *(Stygobromus cooperi)*
- Cascade cave amphipod *(Stygobromus dejectus)*
- Elevated spring amphipod *(Stygobromus elatus)*
- Greenbrier cave amphipod *(Stygobromus emarginatus)*
- Ephemeral cave amphipod *(Stygobromus ephemerus)*
- Ezell's cave amphipod *(Stygobromus flagellatus)*
- Grady's cave amphipod *(Stygobromus gradyi)*
- Devil's sinkhole amphipod *(Stygobromus hadenoecus)*
- Hara's cave amphipod *(Stygobromus harai)*
- Pickle springs amphipod *(Stygobromus heteropodus)*
- Malheur cave amphipod *(Stygobromus hubbsi)*
- Tidewater stygonectid amphipod *(Stygobromus identatus)*
- Long-legged cave amphipod *(Stygobromus longipes)*
- Mackenzie's cave amphipod *(Stygobromus mackenziei)*
- Mountain cave amphipod *(Stygobromus montanus)*
- Morrison's cave amphipod *(Stygobromus morrisoni)*
- Bath County cave amphipod *(Stygobromus mundus)*
- Norton's cave amphipod *(Stygobromus nortoni)*
- Onondaga cave amphipod *(Stygobromus onondagaensis)*
- Ozark cave amphipod *(Stygobromus ozarkensis)*
- Minute cave amphipod *(Stygobromus parvus)*
- Pizzini's amphipod *(Stygobromus pizzinii)*
- Wisconsin well amphipod *(Stygobromus putealis)*

- Reddell's cave amphipod *(Stygobromus reddelli)*
- Alabama well amphipod *(Stygobromus smithii)*
- Spring cave amphipod *(Stygobromus spinatus)*
- Stellmack's cave amphipod *(Stygobromus stellmacki)*
- Subtle cave amphipod *(Stygobromus subtilis)*
- Wengeror's cave amphipod *(Stygobromus wengerorum)*

Niphargids

Species

- Carinurella paradoxa
- Niphargobates lefkodemonaki
- Niphargobates orophobata
- Niphargus aberrans
- Niphargus hadzii
- Niphargus hrabei
- Niphargus sphagnicolus
- Niphargus spoeckeri
- Niphargus stenopus
- Niphargus timavi
- Niphargus valachicus

Subspecies

- Niphargus ilidzensis slovenicus
- Niphargus orcinus orcinus
- Niphargus stygius stygius

Anaspidacea

- *Allanaspides helonomus*
- Hickman's pygmy mountain shrimp *(Allanaspides hickmani)*
- *Eucrenonaspides oinotheke*
- Great Lake shrimp *(Paranaspides lacustris)*

Decapods

There are 224 decapod species and seven decapod subspecies assessed as vulnerable.

Parastacids

- Common yabby *(Cherax destructor)*
- *Cherax papuanus*
- Strzelecki burrowing crayfish *(Engaeus rostrogaleatus)*
- *Engaeus urostrictus*
- Burnie burrowing crayfish *(Engaeus yabbimunna)*
- Glenelg river crayfish *(Euastacus bispinosus)*
- *Euastacus simplex*
- *Euastacus sulcatus*
- New England crayfish *(Euastacus suttoni)*
- *Euastacus yarreansis*
- *Geocharax falcata*
- *Ombrastacoides pulcher*

Gecarcinucids

- Adeleana forcarti
- Arachnothelphusa melanippe
- Austrothelphusa tigrina
- *Austrothelphusa valentula*
- *Heterothelphusa fatum*
- Johnson's freshwater crab *(Irmengardia johnsoni)*
- *Liotelphusa quadrata*
- *Mainitia mainitensis*
- *Mekhongthelphusa kengsaphu*
- *Mekhongthelphusa tetragona*
- *Nautilothelphusa zimmeri*
- *Oziotelphusa biloba*
- *Oziotelphusa hippocastanum*
- *Oziotelphusa ritigala*
- *Oziotelphusa stricta*
- *Oziotelphusa wagrakarowensis*
- *Parathelphusa balabac*
- *Parathelphusa cabayugan*
- *Parathelphusa crocea*
- *Parathelphusa maindroni*
- *Parathelphusa ovum*

- *Parathelphusa pantherina*
- *Parathelphusa possoensis*
- Perbrinckia fenestra
- Perbrinckia integra
- Phricotelphusa callianira
- Phricotelphusa elegans
- Phricotelphusa limula
- Phricotelphusa ranongi
- Sayamia maehongsonensis
- Stygothelphusa bidiensis
- Sundathelphusa minahassae
- Sundathelphusa rubra

Atyids

- Archaetya chacei
- Atya ortmannioides
- Atyaephyra strymonensis
- Caridina ablepsia
- Caridina acuta
- Caridina acutirostris
- Caridina anislaq
- Caridina batuan
- Caridina boholensis
- Caridina breviata
- Caridina bunyonyiensis
- Caridina caerulea
- Caridina camaro
- Caridina caverna
- Caridina cavernicola
- Caridina demenica
- Caridina dianchiensis
- Caridina ensifera
- Caridina feixiana
- Caridina gordonae
- Caridina gortio
- Caridina guangxiensis

- Caridina leclerci
- Caridina longidigita
- Caridina mengae
- Caridina minidentata
- Caridina pseudodenticulata
- Caridina pseudonilotica
- *Caridina samar*
- *Caridina sarasinorum*
- *Caridina schenkeli*
- *Caridina semiblepsia*
- *Caridina sodenensis*
- *Caridina spelunca*
- *Caridina trifasciata*
- *Caridina valencia*
- *Edoneus erwini*
- *Edoneus marulas*
- *Edoneus sketi*
- *Marosina brevirostris*
- *Marosina longirostris*
- *Neocaridina brevidactyla*
- Kentucky cave shrimp *(Palaemonias ganteri)*
- *Paracaridina longispina*
- *Parisia deharvengi*
- *Parisia macrophora*
- *Stygiocaris lancifera*
- *Stygiocaris stylifera*
- Troglocaris bosnica
- Troglocaris kapelana
- Troglocaris neglecta
- Troglocaris prasence
- Typhlatya consobrina
- Typhlatya elenae
- Typhlatya garciadebrasi
- Typhlatya garciai
- Typhlatya miravetensis
- Typhlatya taina

- Typhlocaridina lanceifrons
- Typhlocaridina liui
- Typhlocaridina semityphlata

Cambarids

Species

- Short mountain crayfish *(Cambarus clivosus)*
- Grandfather mountain crayfish *(Cambarus eeseeohensis)*
- Elk river crayfish *(Cambarus elkensis)*
- Alabama cave crayfish *(Cambarus jonesi)*
- Saluda burrowing crayfish *(Distocambarus hunteri)*
- Newberry burrowing crayfish *(Distocambarus youngineri)*
- Daisy burrowing crayfish *(Fallicambarus jeanae)*
- Crittenden crayfish *(Orconectes bisectus)*
- Coldwater crayfish *(Orconectes eupunctus)*
- Yazoo crayfish *(Orconectes hartfieldi)*
- Tennessee cave crayfish *(Orconectes incomptus)*
- Big creek crayfish *(Orconectes peruncus)*
- St. Francis river crayfish *(Orconectes quadruncus)*
- Kiamichi crayfish *(Orconectes saxatilis)*
- Crescent crayfish *(Orconectes taylori)*
- Hardin crayfish *(Orconectes wrighti)*
- Gabriel cave crayfish *(Procambarus cavernicola)*
- *Procambarus citlaltepetl*
- *Procambarus ruthveni*

Subspecies

- Calcasieu crayfish *(Orconectes hathawayi blacki)*
- *Orconectes inermis testii*
- *Orconectes meeki brevis*
- Procambarus lucifugus alachua
- Procambarus rogersi campestris
- Procambarus rogersi ochlocknensis
- Procambarus rogersi rogersi

Potamonautids

- *Boreas uglowi*
- Nimba stream crab *(Liberonautes nimba)*
- *Madagapotamon humberti*
- *Potamonautes choloensis*
- *Potamonautes gerdalensis*
- *Potamonautes ignestii*
- *Potamonautes infravallatus*
- Blue river crab *(Potamonautes lividus)*
- *Potamonautes montivagus*
- *Potamonautes pilosus*
- East African tree hole crab *(Potamonautes raybouldi)*
- Reid's river crab *(Potamonautes reidi)*
- *Potamonautes triangulus*
- *Potamonautes unisulcatus*
- *Potamonautes xiphoidus*
- Sachs's stream crab *(Potamonemus sachsi)*

Pseudothelphusids

- Allacanthos pittieri
- Chaceus ibiricensis
- Elsalvadoria zurstrasseni
- Epilobocera haytensis
- Epilobocera wetherbeei
- Fredius granulatus
- Hypolobocera alata
- Hypolobocera andagoensis
- Hypolobocera barbacensis
- Hypolobocera cajambrensis
- Hypolobocera gracilignatha
- Hypolobocera rathbuni
- Hypolobocera rotundilobata
- Hypolobocera velezi
- Lindacatalina sumacensis
- Microthelphusa forcarti
- Moritschus altaquerensis

- Moritschus ecuadorensis
- Neopseudothelphusa fossor
- Phrygiopilus acanthophallus
- Potamocarcinus hartmanni
- Potamocarcinus roatensis
- Ptychophallus tristani
- Raddaus mertensi
- Rodriguezus trujillensis

Potamids

- *Cryptopotamon anacoluthon*
- *Geothelphusa miyakoensis*
- South Australia crab *(Geothelphusa nanao)*
- Pingtung crab *(Geothelphusa pingtung)*
- Ze concept of crab *(Geothelphusa takuan)*
- Taroko crab *(Geothelphusa taroko)*
- Wang's crab *(Geothelphusa wangi)*
- Wutai crab *(Geothelphusa wutai)*
- *Indochinamon cua*
- *Indochinamon dangi*
- *Indochinamon guttum*
- *Indochinamon mieni*
- *Iomon luangprabangense*
- *Isolapotamon bauense*
- *Johora counsilmani*
- *Johora gapensis*
- *Johora johorensis*
- *Johora thoi*
- *Malayopotamon granulatum*
- Taiwan's South China sea river crab *(Nanhaipotamon formosanum)*
- *Nemoron nomas*
- *Parapotamon spinescens*
- *Potamon bileki*
- *Pupamon phrae*
- *Sinopotamon ebianense*
- *Sinopotamon hanyangense*

- *Stelomon erawanense*
- *Stelomon kanchanaburiense*
- *Stoliczia bella*
- Stoliczia changmanae
- Stoliczia cognata
- Stoliczia goal
- Stoliczia karenae
- Stoliczia kedahensis
- Stoliczia leoi
- Stoliczia pahangensis
- Stoliczia panhai
- Stoliczia perlensis
- Stoliczia tweedei
- Tiwaripotamon edostilus

Palaemonids

- Leptopalaemon gibbosus
- Leptopalaemon gudjangah
- Leptopalaemon magelensis
- Macrobrachium acanthochirus
- Macrobrachium acherontium
- Macrobrachium elegantum
- Macrobrachium gurudeve
- Macrobrachium lingyunense
- Macrobrachium tuxtlaense
- Troglocubanus calcis
- Troglocubanus gibarensis
- Troglocubanus inermis

Trichodactylids

- Avotrichodactylus oaxensis
- Bottiella cucutensis
- Bottiella medemi

Other decapod species

- Noble crayfish *(Astacus astacus)*
- Common spiny lobster *(Palinurus elephas)*
- *Typhlocaris salentina*

Insects

There are 608 insect species and 15 insect subspecies assessed as vulnerable.

Flies

- Belkin's dune tabanid fly *(Brennania belkini)*
- Tasmanian torrent midge *(Edwardsina tasmaniensis)*
- Beautiful-winged fruit fly *(Scaptomyza horaeoptera)*

Plecoptera

- Otway stonefly *(Eusthenia nothofagi)*
- Mount Kosciusko wingless stonefly *(Leptoperla cacuminis)*

Grylloblattaria

- Mount St Helens' grylloblattid *(Grylloblatta chirurgica)*

Orthoptera

There are 137 species in the order Orthoptera assessed as vulnerable.

Crickets

- Howarth's cave cricket *(Caconemobius howarthi)*
- Schauinsland's bush cricket *(Caconemobius schauinslandi)*
- Kaumana cave cricket *(Caconemobius varius)*
- Lanzarote malpais cricket *(Hymenoptila lanzarotensis)*
- *Metioche luteolus*
- Karpathos glandular cricket *(Ovaliptila kinzelbachi)*
- Dodecanese glandular cricket *(Ovaliptila nana)*
- *Phalangacris phaloricephala*
- Volcanoes cave cricket *(Thaumatogryllus cavicola)*
- Kauai thin-footed bush cricket *(Thaumatogryllus variegatus)*

Acridids

- Usambara slant-faced grasshopper *(Acrida bara)*
- Uvarov's bird grasshopper *(Acridoderes uvarovi)*
- Idaho point-headed grasshopper *(Acrolophitus pulchellus)*
- Usambara burrowing grasshopper *(Acrotylus apicalis)*
- Usambara forest edge grasshopper *(Afrophlaeoba usambarica)*
- Michigan bog grasshopper *(Appalachia arcana)*
- Wembere grasshopper *(Aulacobothrus popovi)*
- Tanzanian miombo grasshopper *(Cardeniopsis regalis)*
- Siskiyou Chloealtis grasshopper *(Chloealtis aspasma)*
- White-tipped grasshopper *(Chorthippus acroleucus)*
- Ufipa blue-winged grasshopper *(Chromochokwea fitzgeraldi)*
- Calabrian gold grasshopper *(Chrysochraon beybienkoi)*
- Banat grasshopper *(Eozubovskya banatica)*
- Udzungwa noble grasshopper *(Eupropacris uniformis)*
- *Eximacris superbum*
- Three-spotted forest grasshopper *(Heteracris trimaculata)*
- Reatine Italian grasshopper *(Italohippus modestus)*
- Kilimanjaro drumming grasshopper *(Ixalidium sjostedti)*
- Usambara drumming grasshopper *(Ixalidium transiens)*
- Long-winged mountain grasshopper *(Miramella irena)*
- Romanian mountain grasshopper *(Odontopodisma montana)*
- Red-legged mountain grasshopper *(Odontopodisma rubripes)*
- Erimanthos mountain grasshopper *(Oropodisma erymanthosi)*
- Kyllini mountain grasshopper *(Oropodisma kyllinii)*
- Macedonian mountain grasshopper *(Oropodisma macedonica)*
- Tayetos mountain grasshopper *(Oropodisma taygetosi)*
- Turkish black sea coast grasshopper *(Rammeihippus turcicus)*
- Almeria sand grasshopper *(Sphingonotus almeriense)*
- Tenerife sand grasshopper *(Sphingonotus picteti)*
- Rugose sand grasshopper *(Sphingonotus rugosus)*
- Eurasian toothed grasshopper *(Stenobothrus eurasius)*

Stenopelmatids

- Kelso Jerusalem cricket *(Ammopelmatus kelsoensis)*
- Port Conception Jerusalem cricket *(Ammopelmatus muwu)*

- Poor knights weta *(Deinacrida fallai)*
- Wetapunga *(Deinacrida heteracantha)*
- Stephens Island weta *(Deinacrida rugosa)*
- Coachella Valley Jerusalem cricket *(Stenopelmatus cahuilaensis)*
- Navajo Jerusalem cricket *(Stenopelmatus navajo)*

Tettigoniids

- Richtersveld katydid *(Africariola longicauda)*
- Marakele delicate katydid *(Amyttacta marakelensis)*
- Short-winged tonged bush-cricket *(Anadrymadusa brevipennis)*
- Dilated false shieldback *(Aroegas dilatatus)*
- Flat-necked shieldback *(Arytropteris basalis)*
- *Austrosaga spinifer*
- Sierra Nevadan saddle bush-cricket *(Baetica ustulata)*
- Nihoa conehead katydid *(Banza nihoa)*
- Big pine key conehead katydid *(Belocephalus micanopy)*
- Keys short-winged conehead katydid *(Belocephalus sleighti)*
- Tenerife green bush-cricket *(Calliphona koenigi)*
- Canarian laurel bush-cricket *(Canariola nubigena)*
- Lalande's black-winged clonia *(Clonia lalandei)*
- Uvarov's clonia *(Clonia uvarovi)*
- Peringuey's meadow katydid *(Conocephalus peringueyi)*
- Zlobin's meadow katydid *(Conocephalus zlobini)*
- Remote conehead katydid *(Euconocephalus remotus)*
- Annamaria's marbled bush-cricket *(Eupholidoptera annamariae)*
- Cretan marbled bush-cricket *(Eupholidoptera cretica)*
- Idi marbled bush-cricket *(Eupholidoptera forcipata)*
- Skaronero marbled bush-cricket *(Eupholidoptera gemellata)*
- Giulia's marbled bush-cricket *(Eupholidoptera giuliae)*
- Ikaria marbled bush-cricket *(Eupholidoptera icariensis)*
- Jacqueline's marbled bush-cricket *(Eupholidoptera jacquelinae)*
- Hidden marbled bush-cricket *(Eupholidoptera latens)*
- Lefkas marbled bush-cricket *(Eupholidoptera leucasi)*
- Marianne's marbled bush-cricket *(Eupholidoptera mariannae)*
- Pale-legged marbled bush-cricket *(Eupholidoptera pallipes)*
- *Hemisaga lucifer*

- *Hemisaga vepreculae*
- *Metrioptera domogledi*
- Southern barbed-wire bush-cricket *(Onconotus servillei)*
- *Pachysaga munggai*
- Drakensberg grass false shieldback *(Paracilacris lateralis)*
- Golden Gate grass false shieldback *(Paracilacris mordax)*
- Vardousia Greek bush-cricket *(Parnassiana coracis)*
- Dirphys Greek bush-cricket *(Parnassiana dirphys)*
- Tawny Greek bush-cricket *(Parnassiana fusca)*
- Parnon Greek bush-cricket *(Parnassiana parnon)*
- Slim Greek bush-cricket *(Parnassiana tenuis)*
- *Phasmodes jeeba*
- *Psacadonotus seriatus*
- Gavrogo bush-cricket *(Rhacocleis crypta)*
- Cretan bush-cricket *(Rhacocleis derrai)*
- Lesbos bush-cricket *(Rhacocleis distinguenda)*
- Ferdinand's bush-cricket *(Rhacocleis ferdinandi)*
- Stone-jumping bush-cricket *(Rhacocleis lithoscirtetes)*
- Common predatory bush-cricket *(Saga pedo)*
- Golden Gate seedpod shieldback *(Thoracistus aureoportalis)*
- Green-kneed seedpod shieldback *(Thoracistus viridicrus)*
- *Windbalea viride*
- Zaprochilus ninae

Rhaphidophorids

- Arizona giant sand treader cricket *(Daihinibaenetes arizonensis)*
- Kalymnos cave-cricket *(Dolichopoda calidnae)*
- Cassagnau's cave-cricket *(Dolichopoda cassagnaui)*
- Kefalari cave-cricket *(Dolichopoda dalensi)*
- Lefkas cave-cricket *(Dolichopoda gasparoi)*
- Giuliana's cave-cricket *(Dolichopoda giulianae)*
- Perama cave-cricket *(Dolichopoda graeca)*
- Ithaka cave-cricket *(Dolichopoda ithakii)*
- Kalithea cave-cricket *(Dolichopoda kalithea)*
- Matsakis' cave-cricket *(Dolichopoda matsakisi)*
- Naxos cave-cricket *(Dolichopoda naxia)*

- Cephalonia cave-cricket *(Dolichopoda pavesii)*
- Pan cave-cricket *(Dolichopoda petrochilosi)*
- Thassos cave-cricket *(Dolichopoda thasosensis)*
- Kelso giant sand treader cricket *(Macrobaenetes kelsoensis)*
- Coachelia giant sand treader cricket *(Macrobaenetes valgum)*
- *Pristoceuthophilus sp. nov.*
- *Tasmanoplectron isolatum*
- Tanner's black camel cricket *(Utabaenetes tanneri)*

Phaneropterids

- Mute winter katydid *(Brinckiella aptera)*
- Karoo winter katydid *(Brinckiella karooensis)*
- Mauerberger's winter katydid *(Brinckiella mauerbergerorum)*
- Cozia plump bush-cricket *(Isophya harzi)*
- Long-tailed plump bush-cricket *(Isophya longicaudata)*
- Athos bright bush-cricket *(Poecilimon athos)*
- Ikaria bright bush-cricket *(Poecilimon ikariensis)*
- Istanbul bright bush-cricket *(Poecilimon istanbul)*
- Kadiytsa bright bush-cricket *(Poecilimon pechevi)*
- Short-tailed bull bush-cricket *(Polysarcus scutatus)*
- East coast katydid *(Pomatonota dregii)*

Other Orthoptera species

- Canarian tiny cricket *(Cycloptiloides canariensis)*
- Keys scaly cricket *(Cycloptilum irregularis)*
- Pinaleno monkey grasshopper *(Eumorsea pinaleno)*
- Cretan stone grasshopper *(Orchamus raulinii)*
- Desert monkey grasshopper *(Psychomastax deserticola)*
- Sierra pygmy grasshopper *(Tetrix sierrana)*
- *Uvarovitettix transsylvanicus*

Hymenoptera

There are 155 species in the order Hymenoptera assessed as vulnerable.

Ants

- Acanthomyops latipes

- Acanthomyops murphii
- Anergates atratulus
- Anoplolepis nuptialis
- Antichthonidris bidentatus
- Aphaenogaster bidentatus
- Camponotus universitatis
- Cardiocondyla zoserka
- Cataglyphis hannae
- Chalepoxenus brunneus
- Chalepoxenus kutteri
- Chalepoxenus muellerianus
- Chalepoxenus spinosus
- Chalepoxenus tarbinskii
- Chalepoxenus tauricus
- Chalepoxenus tramieri
- Chalepoxenus zabelini
- Crematogaster atilanica
- Crematogaster pilosa
- Doronomyrmex goesswaldi
- Doronomyrmex kutteri
- Doronomyrmex pacis
- Doronomyrmex pocahontas
- Dorymyrmex insanus
- Epimyrma adlerzi
- Epimyrma africana
- Epimyrma algeriana
- Epimyrma bernardi
- Epimyrma corsica
- Epimyrma goridaghini
- Epimyrma kraussei
- Ravoux's slavemaker ant *(Epimyrma ravouxi)*
- *Epimyrma stumperi*
- *Epimyrma tamarae*
- *Epimyrma zaleskyi*
- *Formica dirksi*
- *Formica talbotae*

- *Formicoxenus chamberlini*
- *Formicoxenus nitidulus*
- *Formicoxenus provancheri*
- *Formicoxenus quebecensis*
- *Formicoxenus sibiricus*
- *Harpagoxenus canadensis*
- *Harpagoxenus sublaevis*
- *Harpagoxenus zaisanicus*
- *Kyidris media*
- *Kyidris yaleogyna*
- *Lasius reginae*
- *Leptothorax buschingeri*
- *Leptothorax duloticus*
- *Leptothorax faberi*
- Leptothorax minutissimus
- Manica parasitica
- Megalomyrmex symmetochus
- Monomorium effractor
- Monomorium hospitum
- Monomorium inquilinum
- Monomorium noualhieri
- Monomorium pergandei
- Monomorium santschii
- Monomorium talbotae
- Myrmecia inquilina
- Myrmica bibikoffi
- Myrmica colax
- Myrmica ereptrix
- Myrmica faniensis
- Myrmica hirsuta
- Myrmica kabylica
- Myrmica lampra
- Myrmica laurae
- Myrmica lemasnei
- Myrmica myrmicoxena
- Myrmica quebecensis

- Myrmica samnitica
- Myrmica symbiotica
- Myrmica winterae
- Myrmoxenus gordiagini
- Oxyepoecus bruchi
- Oxyepoecus daguerrei
- Oxyepoecus inquilinus
- Pheidole acutidens
- Pheidole argentina
- Pheidole elecebra
- Pheidole inquilina
- Pheidole lanuginosa
- Pheidole microgyna
- Pheidole neokohli
- Pheidole oculata
- Pheidole parasitica
- Pheidole symbiotica
- Plagiolepis ampeloni
- Plagiolepis grassei
- Plagiolepis regis
- Pogonomyrmex anergismus
- Pogonomyrmex colei
- Polyergus breviceps
- Polyergus lucidus
- Polyergus nigerrimus
- Polyergus samurai
- Protomognathus americanus
- Pseudoatta argentina
- Pseudomyrmex leptosus
- Rhoptromyrmex mayri
- Rhoptromyrmex schmitzi
- Rossomyrmex minuchae
- Rossomyrmex proformicarum
- Serrastruma inquilina
- Solenopsis daguerrei
- Solenopsis solenopsidis

- Strongylognathus afer
- Strongylognathus alboini
- *Strongylognathus alpinus*
- *Strongylognathus arnoldii*
- *Strongylognathus caeciliae*
- *Strongylognathus cecconii*
- *Strongylognathus chelifer*
- *Strongylognathus christophi*
- *Strongylognathus dalmaticus*
- *Strongylognathus destefanii*
- *Strongylognathus emeryi*
- *Strongylognathus foreli*
- *Strongylognathus huberi*
- *Strongylognathus insularis*
- *Strongylognathus italicus*
- Strongylognathus karawajewi *(Strongylognathus karawajevi)*
- *Strongylognathus kervillei*
- *Strongylognathus koreanus*
- Strongylognathus kratochvili *(Strongylognathus kratochvilli)*
- *Strongylognathus minutus*
- *Strongylognathus palaestinensis*
- *Strongylognathus pisarskii*
- Strongylognathus rehbinderi
- Strongylognathus ruzskyi
- Strongylognathus silvestrii
- Strumigenys xenos
- Teleutomyrmex kutteri
- Teleutomyrmex schneideri
- Tetramorium microgyna
- Tetramorium parasiticum

Colletids

- Colletes dimidiatus
- Colletes moricei
- Colletes pulchellus

Melittids

- Melitta hispanica
- Melitta kastiliensis

Apids

- Bombus alpinus
- Obscure bumble bee *(Bombus caliginosus)*
- Yellow bumblebee *(Bombus fervidus)*
- *Bombus medius*
- *Bombus mexicanus*
- Morrison bumble bee *(Bombus morrisoni)*
- Western bumblebee *(Bombus occidentalis)*
- American bumblebee *(Bombus pensylvanicus)*
- Yellow-banded bumblebee *(Bombus terricola)*
- *Nomada noskiewiczi*

Megachilids

- Wallace's giant bee *(Megachile pluto)*

Mantises

- Delicate dwarf mantis *(Ameles gracilis)*
- Fringed dwarf mantis *(Ameles limbata)*
- Short-winged dwarf mantis *(Pseudoyersinia subaptera)*

Lepidoptera

Lepidoptera comprises moths and butterflies. There are 128 species and ten subspecies in the order Lepidoptera assessed as vulnerable.

Lasiocampids

- Small lappet moth *(Phyllodesma ilicifolia)*

Swallowtail butterflies

- *Atrophaneura atropos*
- *Atrophaneura luchti*
- *Atrophaneura schadenbergi*

- Zetides swallowtail *(Battus zetides)*
- Ludlow's Bhutan swallowtail *(Bhutanitis ludlowi)*
- Yellow kite swallowtail *(Eurytides iphitas)*
- Andamans swordtail *(Graphium epaminondas)*
- *Graphium idaeoides*
- Meek's graphium *(Graphium meeki)*
- *Graphium megaera*
- *Graphium procles*
- *Graphium stresemanni*
- Obi Island birdwing *(Ornithoptera aesacus)*
- Rothschild's birdwing *(Ornithoptera rothschildi)*
- *Papilio carolinensis*
- *Papilio esperanza*
- Jordan's swallowtail *(Papilio jordani)*
- Cream-banded swallowtail *(Papilio leucotaenia)*
- *Papilio mangoura*
- *Papilio neumoegeni*
- *Papilio osmana*
- Papillon la pature *(Papilio phorbanta)*
- Fluminense swallowtail *(Parides ascanius)*
- Apollo *(Parnassius apollo)*
- *Parnassius autocrator*
- Jamaican kite *(Protographium marcellinus)*
- Talaud black birdwing *(Troides dohertyi)*
- Buru opalescent birdwing *(Troides prattorum)*
- Caucasian festoon *(Zerynthia caucasica)*

Lycaenids

- Alaena margaritacea
- Aloeides caledoni
- Aloeides carolynnae
- Aloeides dentatis
- *Aloeides kaplani*
- *Aloeides lutescens*
- *Aloeides merces*
- *Aloeides pringlei*

- *Aloeides rossouwi*
- *Aslauga australis*
- *Capys penningtoni*
- *Cyanophrys bertha*
- *Erikssonia acraeina*
- White spotted sapphire *(Iolaus lulua)*
- *Lepidochrysops balli*
- *Lepidochrysops jefferyi*
- *Lepidochrysops littoralis*
- *Lepidochrysops loewensteini*
- *Lepidochrysops oosthuizeni*
- *Lepidochrysops outeniqua*
- *Lepidochrysops pephredo*
- *Lepidochrysops poseidon*
- *Lepidochrysops pringlei*
- *Lepidochrysops swanepoeli*
- *Lepidochrysops titei*
- *Lepidochrysops victori*
- *Lepidochrysops wykehami*
- Tiassale liptena *(Liptena tiassale)*
- Hermes copper *(Lycaena hermes)*
- Grecian copper *(Lycaena ottomana)*
- *Orachrysops ariadne*
- *Phasis pringlei*
- *Poecilmitis adonis*
- Azure opal *(Poecilmitis azurius)*
- *Poecilmitis balli*
- *Poecilmitis daphne*
- *Poecilmitis endymion*
- *Poecilmitis henningi*
- *Poecilmitis hyperion*
- *Poecilmitis irene*
- *Poecilmitis kaplani*
- Tsomo River copper *(Poecilmitis lyncurium)*
- *Poecilmitis lyndseyae*
- *Poecilmitis orientalis*

- *Poecilmitis penningtoni*
- *Poecilmitis pyramus*
- *Poecilmitis stepheni*
- *Poecilmitis trimeni*
- *Poecilmitis wykehami*
- Higgin's anomalous blue *(Polyommatus galloi)*
- Sierra Nevada blue *(Polyommatus golgus)*
- Ipharcamon blue *(Polyommatus iphicarmon)*
- Lycian blue *(Polyommatus lycius)*
- *Polyommatus orphicus*
- Sinai hairstreak *(Satyrium jebelia)*
- Avalon hairstreak *(Strymon avalona)*
- *Thestor compassbergae*
- *Thestor dryburghi*
- Kaplan's Thestor *(Thestor kaplani)*
- *Thestor pringlei*
- *Thestor rossouwi*
- *Thestor stepheni*
- *Thestor strutti*
- *Thestor swanepoeli*
- Thestor tempe
- Thestor yildizae

Nymphalids

Species

- Madagascan friar *(Amauris nossima)*
- Mauritian friar *(Amauris phoedon)*
- Uncompahgre fritillary *(Boloria acrocnema)*
- Balkan heath *(Coenonympha orientalis)*
- Raetzer's ringlet *(Erebia christi)*
- Sudeten ringlet *(Erebia sudetica)*
- Andaman crow *(Euploea andamanensis)*
- Cordelia's crow *(Euploea cordelia)*
- Mascarene crow *(Euploea euphon)*
- Spartan crow *(Euploea lacon)*

- Magou *(Euploea magou)*
- Nicobar crow *(Euploea scherzeri)*
- El Hierro grayling *(Hipparchia bacchus)*
- La Palma grayling *(Hipparchia tilosi)*
- Electra's tree-nymph *(Idea electra)*
- Bedford-Russell's tree-nymph *(Idea tambusisiana)*
- *Ideopsis oberthurii*
- New Ireland yellow tiger *(Parantica clinias)*
- D'Abrera's tiger *(Parantica dabrerai)*
- Dannatt's tiger *(Parantica dannatti)*
- Angled tiger *(Parantica garamantis)*
- Sumbawa tiger *(Parantica philo)*
- Felder's tiger *(Parantica phyle)*
- Toxopeus' yellow tiger *(Parantica toxopei)*
- Flores tiger *(Parantica wegneri)*
- Dils' grayling *(Pseudochazara orestes)*
- Crow tiger *(Tirumala euploeomorpha)*

Subspecies

- *Danaus affinis jimiensis*
- *Danaus melanippus keteus*
- Outcast crow *(Euploea algea abjecta)*
- *Euploea algea eleutho*
- *Euploea algea schmeltzi*
- *Euploea crameri albomaculata*
- *Euploea mulciber elwesii*
- *Euploea radamanthus schreiberi*
- *Parantica weiskei thalassina*
- Sarangani tiger *(Tirumala choaspes tumanana)*

Skippers

- Dakota Skipper *(Hesperia dacotae)*
- Cinquefoil skipper *(Pyrgus cirsii)*

Pierids

- Canary brimstone *(Gonepteryx cleobule)*

- Atewa dotted border *(Mylothris atewa)*
- Moroccan green-veined white *(Pieris segonzaci)*

Beetles

There are 50 beetle species assessed as vulnerable.

Dytiscids

- Acilius duvergeri
- Deronectes depressicollis
- Deronectes ferrugineus
- Dytiscus latissimus
- Graphoderus bilineatus
- Meladema lanio

Longhorn beetles

- Cerambyx longicorn *(Cerambyx cerdo)*
- *Clytus clavicornis*
- *Clytus triangulimacula*
- *Delagrangeus schurmanni*
- *Isotomus barbarae*
- *Macrodontia cervicornis*
- *Morimus funereus*
- Rosalia longicorn *(Rosalia alpina)*

Scarabaeids

- Ciervo scarab beetle *(Aegialia concinna)*
- Cresent dune scarab beetle *(Aegialia crescenta)*
- *Ateuchetus semipunctatus*
- *Ateuchus squalidus*
- *Aulacopris matthewsi*
- *Canthon corpulentus*
- *Canthon quadripunctatus*
- *Clypeodrepanus striatus*
- *Ontherus hadros*
- *Onthophagus albarracinus*
- *Pachysoma aesculapius*

- *Pachysoma endroedyi*
- *Pachysoma glentoni*
- *Pedaridium hirsutum*
- *Platyonitis bicuariensis*
- Giuliani's dune scarab beetle *(Pseudocotalpa giulianii)*

Other beetle species

- *Ampedus brunnicornis*
- *Ampedus hjorti*
- *Anonyxmolytes lilliput*
- *Carabus olympiae*
- Cobblestone tiger beetle *(Cicindela marginipennis)*
- Globose dune beetle *(Coelus globosus)*
- San Joaquin dune beetle *(Coelus gracilis)*
- *Colophon cameroni*
- *Colophon neli*
- *Colophon stokoei*
- *Colophon westwoodi*
- Blind cave beetle *(Glacicavicola bathysciodes)*
- *Gnorimus decempunctatus*
- *Leipaspis lauricola*
- *Melasis fermini*
- *Microblattellus lecongmani*
- Frigate Island giant tenebrionid beetle *(Polposipus herculeanus)*
- Gammon's riffle beetle *(Stenelmis gammoni)*
- *Thorectes valencianus*
- *Triplax emgei*

Odonata

Odonata includes dragonflies and damselflies. There are 129 species and five subspecies in the order Odonata assessed as vulnerable.

Chlorogomphids

- Chlorogomphus brevistigma
- Chlorogomphus gracilis
- Chlorogomphus nakamurai

- Chlorogomphus xanthoptera
- Chloropetalia selysi
- Watanabeopetalia uenoi

Chlorocyphids

- Banded jewel *(Africocypha centripunctata)*
- Kaleidoscope jewel *(Africocypha lacuselephantum)*
- Albertine jewel *(Chlorocypha schmidti)*
- *Libellago andamanensis*
- Tanzania jewel *(Platycypha auripes)*
- *Rhinocypha dorsosanguinea*
- *Rhinocypha latimacula*
- Mahale jewel *(Stenocypha hasta)*

Platycnemidids

- Calicnemia nipalica
- Coeliccia exoleta
- Coeliccia flavostriata
- Coeliccia fraseri
- Disparoneura apicalis
- Yellow-fronted threadtail *(Elattoneura dorsalis)*
- *Elattoneura pasquinii*
- *Lieftinckia lairdi*
- *Nososticta phoenissa*
- *Nososticta plagioxantha*
- *Risiocnemis pulchra*
- Ceres stream-damsel *(Spesbona angusta)*

Megapodagrionids

- *Heteragrion flavovittatum*
- Cuban hypolestes *(Hypolestes trinitatis)*
- Zoe waterfall damsel *(Paraphlebia zoe)*
- *Philogenia cristalina*
- *Philosina alba*
- Little wood elf *(Sciotropis cyclanthorum)*
- *Teinopodagrion temporale*

Gomphids

Species

- Top end dragon *(Antipodogomphus dentosus)*
- Cape thorntail *(Ceratogomphus triceraticus)*
- *Crenigomphus abyssinicus*
- *Crenigomphus denticulatus*
- *Cyclogomphus gynostylus*
- *Gomphus lucasii*
- Tennessee clubtail *(Gomphus sandrius)*
- Westfall's clubtail *(Gomphus westfalli)*
- Kakadu vicetail *(Hemigomphus magela)*
- *Lamelligomphus tutulus*
- *Macrogomphus lankanensis*
- *Nepogomphoides stuhlmanni*
- Dark pincertail *(Onychogomphus assimilis)*
- Waved pincertail *(Onychogomphus flexuosus)*
- Levant pincertail *(Onychogomphus macrodon)*
- *Paragomphus tachyerges*
- Pegtail *(Perigomphus pallidistylus)*
- *Phyllogomphoides joaquini*
- Yellow-sided clubtail *(Stylurus potulentus)*
- Townes's clubtail *(Stylurus townesi)*

Subspecies

- Erpetogomphus lampropeltis lampropeltis

Coenagrionids

Species

- Chimanimani bluet *(Africallagma cuneistigma)*
- Liberian wisp *(Agriocnemis angustirami)*
- *Antiagrion blanchardi*
- Sabino dancer *(Argia sabino)*
- Yellow waxtail *(Ceriagrion citrinum)*
- Turkish red damsel *(Ceriagrion georgifreyi)*

- Cretan bluet *(Coenagrion intermedium)*
- East coast giant *(Coryphagrion grandis)*
- Cuban bluet *(Enallagma truncatum)*
- Ethiopian bluetail *(Ischnura abyssinica)*
- San Francisco forktail *(Ischnura gemina)*
- *Ischnura pamelae*
- *Mecistogaster asticta*
- Oahu damselfly *(Megalagrion oahuense)*
- Oceanic Hawaiian damselfly *(Megalagrion oceanicum)*
- Pacific Hawaiian damselfly *(Megalagrion pacificum)*
- Orangeblack Hawaiian damselfly *(Megalagrion xanthomelas)*
- *Mesamphiagrion ovigerum*
- *Mesamphiagrion rosseri*
- *Pericnemis triangularis*
- *Pseudagrion bicoerulans*
- *Pseudagrion guichardi*
- *Pseudagrion kaffinum*
- Harlequin sprite *(Pseudagrion newtoni)*
- *Pseudagrion pontogenes*
- Vumba sprite *(Pseudagrion vumbaense)*
- Indian Ocean fineliner *(Teinobasis alluaudi)*
- *Telebasis farcimentum*

Subspecies

- Somalian bluet *(Azuragrion somalicum amitinum)*
- Frey's damselfly *(Coenagrion hylas freyi)*
- Blackline damselfly *(Megalagrion nigrohamatum nigrolineatum)*
- *Pseudagrion sublacteum mortoni*

Aeshnids

- Thylacine darner *(Acanthaeschna victoria)*
- *Gynacantha bispina*
- *Gynacantha constricta*
- *Petaliaeschna flavipes*
- Yemen hawker *(Pinheyschna yemenensis)*
- *Planaeschna celia*

- *Sarasaeschna niisatoi*

Libellulids

- *Atoconeura aethiopica*
- *Bironides teuchestes*
- Dark-winged groundling *(Brachythemis fuscopalliata)*
- *Diplacina arsinoe*
- Purple skimmer *(Libellula jesseana)*
- *Micrathyria divergens*
- *Micrathyria pseudhypodidyma*
- Mangrove skimmer *(Orthetrum poecilops)*
- *Sympetrum gracile*
- *Urothemis abbotti*

Other Odonata species

- Argiolestes realensis
- *Bayadera hyalina*
- Elegant malachite *(Chlorolestes elegans)*
- *Cora lugubris*
- Say's spiketail *(Cordulegaster sayi)*
- *Drepanosticta centrosaurus*
- Marbled malachite *(Ecchlorolestes peringueyi)*
- *Euphaea basalis*
- *Euphaea pahyapi*
- Guatemalan rubyspot *(Hetaerina rudis)*
- Rock narrow-wing *(Lithosticta macra)*
- *Macromia erato*
- *Macromia irina*
- *Macromia katae*
- Splendid cruiser *(Macromia splendens)*
- *Metaphya elongata*
- African twigtail *(Nubiolestes diotima)*
- Red relic *(Pentaphlebia stahli)*
- Peaked redspot *(Phyllopetalia excrescens)*
- *Platysticta deccanensis*
- *Protosticta khaosoidaoensis*

- *Protosticta sanguinostigma*
- Bulgarian emerald *(Somatochlora borisi)*
- Texas emerald *(Somatochlora margarita)*
- Yellow presba *(Syncordulia gracilis)*
- Mahogany presba *(Syncordulia venator)*
- *Synthemis alecto*
- Green-banded sparklewing *(Umma declivium)*
- Angola sparklewing *(Umma femina)*
- Purple sparklewing *(Umma purpurea)*
- Ringed boghaunter *(Williamsonia lintneri)*

MARINE CONSERVATION ACTIVISM

Marine Conservation Activism refers to the efforts of non-governmental organizations and individuals to bring about social and political change in the area of marine conservation. Marine conservation is properly conceived as a set of management strategies for the protection and preservation of ecosystems in oceans and seas. Activists raise public awareness and support for conservation, while pushing governments and corporations to practice sound ocean management, create conservation policy, and enforce existing laws and policy through effective regulation. There are many different kinds of organizations and agencies that work toward these common goals. They all are a part of the growing movement that is ocean conservation. These organizations fight for many causes including stopping pollution, overfishing, whaling, by-catching, and Marine Protected Areas.

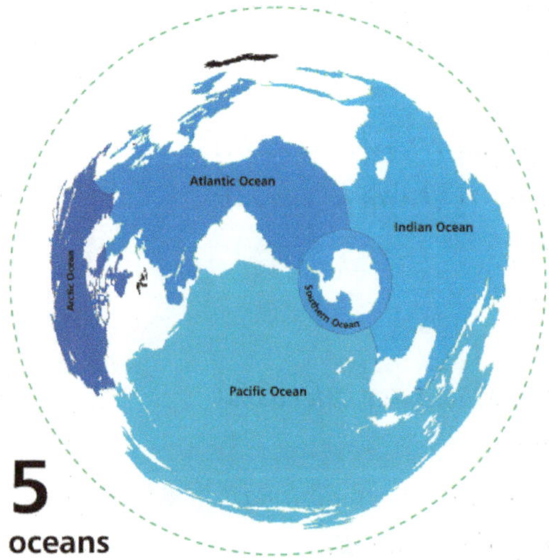

Animated map of the world's oceanic waters.

History

United States

Though the environmental movement began in the United States during the 1960s, the idea of marine conservation really did not take off in the country until the 1972 when the Marine Protection, Research, and Sanctuaries Act (MPRSA) passed, beginning the movement. The act allowed the regulation by the United States Environmental Protection Agency (EPA) over dumping in the seas. Though the act was later amended, it was one of several key events to bring marine issues towards the front of environmental issues in the United States.

Notable People

- Jacques Cousteau: Explorer, Conservationist, Researcher & Author
- Sylvia Earle: Marine Biologist, Explorer, & Author
- Paul Watson: Conservationist & Activist
- Steve Irwin: Naturalist, Conservationist, Zoologist, Herpetologist, & Television Personality

International Issues

Debris

Marine debris on the Hawaiian coast

Marine Debris is defined as "any persistent solid material that is manufactured or processed and directly or indirectly, intentionally or unintentionally, disposed of or abandoned into the marine environment or the Great Lakes". This debris can injure or even kill marine organisms; it can also interfere with navigation safety and could pose a threat to human health. Marine debris can range from soda cans to plastic bags and can even include abandoned vessels or neglected fishing gear.

Ocean Conservancy

Ocean Conservancy is a non-profit environmental group that fights for the improvement and conservation of marine ecosystems and marine life. They work to find science-based solutions to protect the world's oceans from the global challenges that they face today.

One of the many issues that they work closely to stop is the flow of trash that enters the ocean. The International Coastal Cleanup (ICC) is one of the methods Ocean Conservancy uses to prevent marine debris. The ICC is the largest volunteer effort to clean up the world›s oceans and other waterways; over the past 25 years the ICC has cleaned up approximately 144,606,491 pounds of trash from beaches all over the world.

Whaling

Whaling in the Faroe Islands.

International Whaling Commission

Whaling is the hunting of free roaming whales; many whaling practices have led to drastic population loss in many whale populations around the world. In 1986, The International Whaling Commission (IWC) was founded to put a ban on commercial whaling. The commission recognizes three different types of whaling: aboriginal subsistence, commercial, and special permit (or scientific) whaling.

Aboriginal Subsistence Whaling

This form of whaling supports indigenous communities where whale products play an important role in cultural and nutritional life. The IWC sets catch limits for aboriginal subsistence whaling every six years.

Commercial Whaling

This form of whaling is highly regulated by the IWC and is currently on a moratorium. There are a few countries that oppose the moratorium and continue to hunt for whales; these countries share catch data with the Commission but are not regulated by it.

Since the moratorium was put in place in 1986, more than 50,000 whales have been hunted and killed; there are three nations that are still able to hunt whales because of loopholes in the ban. Norway is able to hunt because of an "objection" to the ban; Iceland is able to hunt because of a "reservation" and Japan is able to hunt because they claim it is for "research purposes". If combined these nations kill around 2,000 whales each year; these whales include humpback, minke, sperm, fin, Bryde's, and sei. The IWC ban does allow for some Aboriginal Subsistence Whaling (ASW) in certain countries.

Special Permit/Scientific Whaling

This category of whaling is separated from IWC-regulated whaling by international law. Special permit research proposals are to be submitted by countries to the IWC for scientific scrutiny. The role of the IWC is advisory only.

Greenpeace

Greenpeace, an international environmental organization founded in 1971 in British Columbia, fights against whaling. Their campaigns are nonviolent and many times involve one or more of the five Greenpeace ships which first made the organization famous in the 1970s. In late December 2005, Japanese whaling fleets experienced heavy opposition from Greenpeace, who protested that the Japanese were continuing their commercial whaling under the guise of research, and even worse, they were doing so in the Southern Ocean Whale Sanctuary. They sent volunteer workers in inflatable boats to get in the line of fire in order to stop the whaling.

Sea Shepherd Conservation Society

Sea Shepherd Conservation Society is a non-profit, marine wildlife conservation organization that works internationally on numerous campaigns to protect the world's oceans. Their mission is to conserve and protect the world's ecosystems and species; they work to end the destruction of habitat and slaughter of the ocean's wildlife. Unlike many other non-profit environmental groups, Sea Shepherd uses direct-action tactics to expose and challenge illegal activities at sea; they strive to ensure that the ocean can survive for future generations. In doing so, they refer to the United Nations World Charter for Nature that calls on individuals to «safeguard and conserve nature in areas beyond national jurisdiction».

Sea Shepherd was founded in 1977 by Captain Paul Watson in Vancouver BC, Canada; it wasn't until 1981 that it was formally incorporated in the United States. Throughout the years their campaigns have ranged from stopping the annual killing of baby harp seals in Eastern Canada to preventing Japanese whalers from killing endangered whale species. They only work to uphold international conservation law and to protect the endangered ocean habitats and species; they do this without prejudice against race, nationality, color, or religious belief. Their crews are made up of volunteers from all over the world, some of which are from countries that Sea Shepherd has campaigns against; they describe themselves as "pro-ocean" instead of "anti-any nationality or culture".

Shark Finning

Shark finning is a worldwide issue that involves cutting off the fins of sharks. This is done while the shark is still alive followed by the rest of the body being thrown back into the ocean, leaving it to die days after. Used in countries like China and Japan, shark fins are a key ingredient in the world-renowned meal, shark fin soup. The high demand for this particular type of soup has sky rocketed in the last few decades and sells for around $100 on average and is often catered at special occasions such as weddings and banquets. Due to the increased want for these shark fins, traders seek out the fins in order to make a profit. However, the fins are the only part of the shark that fishermen seek out to retrieve due to the low economical value of the actual shark meet. This recently exposed issue along with other overfishing issues has brought upon roughly 80 percent of the shark population decline. It has become prominent concern in marine conservation activism for millions of sharks are killed yearly at an often-unregulated expense.

Project AWARE

Current campaign known as Project AWARE is working globally to advocate solutions for long-term protection for these animals. Created initially as an environmental initiative project, this campaign was developed by the Professional Associations of Diving Instructors (PADI) in 1989. Used to educate divers about environmental problems this program eventually grew to become a registered non-profit organization in the US in1992 and eventually became recognized in the UK and Australia in 1999 and 2002 respectively. In spite of the arising issues with marine challenges, Project AWARE has continued to grow towards meeting the needs of the marine ecosystem as they see fit. Marine debris and shark and ray conservation activism are the two most prevalent issues that are being further worked toward improving since 2011.

Shark Savers

Another huge campaign working to ensure the protection of these marine species is a group called Shark Savers that is sponsored by the group called WildAid. Through the use of community motivation, the project encourages the public to stop eating sharks and shark fin soup. By also working to improve global regulations and creating sanctuaries for sharks, the project aims to take action and get results. Similarly to Project Aware, the Shark Savers program was founded by a group of divers that wanted to help the marine system in 2007. Through the recent creation of shark sanctuaries, the program focuses on sustainability when thinking about the economical and environmental benefits. These created sanctuaries provide a protected area for the sharks and also promote change in nearby communities.

Bite-Back

Bite-Back is another organization that is active in the community and aims to stop the sale of shark fins for the making of shark fin soup in Great Britain. By exposing the UK and their acts toward profiting from shark products, they aim to put and end to their ways of over fishing and exploitation. The organizations main goal is to allow marine life a chance to thrive while they are busy doing the dirty work of lowering consumer stipulation.

Shark Trust

Part of a world-wide alliance called The Global Shark and Ray Initiative (GSRI), the Shark Trust is working in efforts to better the ocean for marine animals such as shark and rays. The Initiative created a plan for changing the status of the shark population that would span over 10 years starting on February 15, 2016. Teaming up with other large conservation organization such as Shark Advocates International and World Wide Fund for Nature (WWF), the team aims to ultimately give these vulnerable animals the safety and security that the ought to have in their natural environment.

Overfishing

Overfishing occurs when fish stocks are over-exploited to below acceptable levels; eventually the fish populations will no longer be able to sustain themselves. This can lead to resource depletion, reduced biological growth, and low biomass levels.

Global Fishing Watch

In September 2016, a partnership of Google and Oceana and Skytruth introduced Global Fishing Watch, a website designed to assist citizens of the globe in monitoring fishing activities.

Solutions

Marine Protected Areas

Even though the idea of Marine Protected Areas is an internationally known concept, there is no one term used internationally. Rather, each country has its own name for the areas. Marine Reserves, Specially Protected Areas, Marine Park all relate to this concept, though differ slightly. Some of the most famous Marine Protected Areas are the Ligurian Sea Cetacean Sanctuary along the coasts of Spain, Monaco, Italy, and Australia's Great Barrier Reef. The largest sanctuary in the world is the Northwestern Hawaiian Islands National Monument. The purpose of these sanctuaries is to provide protection for the living and non- living resources of the oceans and seas. They are created to save species, nursing resources and to help sustain the fish population.

The activists at the Ocean Conservancy fight for this cause. They believe that the United States should put forth a consistent and firm commitment in using Marine Protected Areas as a management strategy. Currently, the argument in the United States is whether or not they are necessary, when it should be how can they work the most efficiently. Activists at the Ocean Conservancy have been working on a campaign called the *Save Our Ocean Legacy,* a campaign lasting several years trying to establish Marine Protected Areas' off California`s coasts. 29 Marine Protected Areas were planned to be established when the legislation bill passed in 1999. The hope is that the plan will be finalized in 2007.

Some fishers do not accept that Marine Protected Areas (MPA) benefit fish stocks and provide insurance against stock collapse. MPAs can cause a short-term loss in fisheries production. However, the concept of spillover, where fish within a Marine Protected Areas (MPAs) move into fished areas, thus benefiting fisheries, has been misunderstood by some fishers.The term is a simplification of numerous ecological benefits that are derived from removing fishing from nursery, breeding grounds and essential fish habitats.

REFERENCES

1. Ray, G. Carleton. "Issues and Mechanisms." Coastal-marine Conservation: Science and Policy. Malden, MA: Blackwell Pub.,2004.
2. "IMPORTANCE OF CORAL REEFS - Coral Reefs - Ocean World". tamu.edu.
3. Trist, Carolyn. "Recreating Ocean Space: Recreational Consumption and Representation of the Caribbean Marine." Professional Geographer. 51.3 (1999). Print.
4. Ngoc, Ninh Thi; Huong, Pham Thi Mai; Thanh, Nguyen Van; Cuong, Nguyen Xuan; Nam, Nguyen Hoai; Thung, Do Cong; Kiem, Phan Van; Minh, Chau Van (2016-09-01). "Steroid Constituents from the Soft Coral Sinularia nanolobata". Chemical & Pharmaceutical Bulletin. 64 (9): 1417–1419. doi:10.1248/cpb.c16-00385. ISSN 1347-5223. PMID 27321426.
5. Yin, Chen-Ting; Wen, Zhi-Hong; Lan, Yu-Hsuan; Chang, Yu-Chia; Wu, Yang-Chang; Sung, Ping-Jyun (2015-01-01). "New Anti-inflammatory Norcembranoids from the Soft Coral Sinularia numerosa". Chemical & Pharmaceutical Bulletin. 63 (9): 752–756. doi:10.1248/cpb.c15-00414. ISSN 1347-5223. PMID 26329871.
6. Burke, Lauretta, Liz Selig, and Mark Spalding. "Reefs At Risk in Southeast Asia." World Resources Institute (2002): 72. Print.
7. J. M. Pandolfi et al., 2003. Global Trajectories of long-term decline of coral reef ecosystems. Science 301: 955-958.
8. "Coral reef destruction and conservation", 18 May 2011
9. Rodrigo, Raul. Resource at Risk: Philippine Coral Reefs. Manila?: Bookmark, 1998.
10. World Resources Institute Southeast Asia Key Findings
11. "FIS - Worldnews - Seafood producers tackle ocean acidification at annual Summit". fis.com.

12. "Is It Time to Sell Your SUV? A Tale of Coral Reef Decline by Eric Borneman - Reefkeeping.com". reefkeeping.com.
13. Archived 10 January 2011 at the Wayback Machine.
14. L. Jenkins. "Profile and Influence of the Successful Fisher-Inventor of Marine Conservation Technology". academia.edu.
15. Thys, Tierney (30 November 2003). "Tracking Ocean Sunfish, Mola mola with Pop-Up Satellite Archival Tags in California Waters". OceanSunfish.org. Retrieved 14 June 2007.
16. Block, Barbara A.; Dewar, Heidi; Farwell, Charles; Prince, Eric D. (1998-08-04). "A new satellite technology for tracking the movements of Atlantic bluefin tuna". Proceedings of the National Academy of Sciences of the United States of America. 95 (16): 9384–9389. doi:10.1073/pnas.95.16.9384. ISSN 0027-8424. PMC 21347Freely accessible. PMID 9689089.
17. Hoolihan, John P. (2004-11-16). "Horizontal and vertical movements of sailfish (Istiophorus platypterus) in the Arabian Gulf, determined by ultrasonic and pop-up satellite tagging". Marine Biology. 146 (5): 1015–1029. doi:10.1007/s00227-004-1488-2. ISSN 0025-3162.
18. Stokesbury, Michael J. W.; Harvey-Clark, Chris; Gallant, Jeffrey; Block, Barbara A.; Myers, Ransom A. (2005-07-21). "Movement and environmental preferences of Greenland sharks (Somniosus microcephalus) electronically tagged in the St. Lawrence Estuary, Canada". Marine Biology. 148 (1): 159 165. doi:10.1007/s00227-005-0061-y. ISSN 0025-3162.
19. "Turtle Excluder Devices". noaa.gov.
20. Turtle Excluder Device on YouTube
21. Techno-Arrogance and Halfway Technologies: Salmon Hatcheries on the Pac c Coast of North America
22. "Ending Commercial Whaling." 19 February 2015.
23. Carleton, Ray G.; McCormick, Jerry (1 April 2009). Coastal-Marine Conservation: Science and Policy. John Wiley & Sons. ISBN 978-1-4443-1124-2.
24. Noakes, Scott. "Atlantic Gray Whale: Research: Science: Gray's Reef National Marine Sanctuary" 19 February 2015.
25. "Hawaiian Monk Seal (Neomonachus Schauinslandi)" NOAA Fisheries. 15 January 2016. 12 February 2015.
26. "Caribbean Monk Seal Gone Extinct From Human Causes, NOAA Confirms" ScienceDaily, 9 June 2008. 12 February 2015.
27. Aidan Bodeo-Lomicky (4 February 2015). The Vaquita: The Biology of an Endangered Porpoise. Createspace Independent Pub. ISBN 978-1-5077-5577-8.
28. Bodeo-Lomicky, Aidan. The Vaquita: The Biology of an Endangered Porpoise. CreateSpace Independent Platform, 2013. Print.
29. "Vaquita" World Wildlife Fund. 12 February 2015.
30. "Green Turtle (Chelonia Mydas)" NOAA Fisheries. 19 February 2015.

31. "Kemp's Ridley Turtle (Lepidochelys Kempii)." : NOAA Fisheries. 19 February 2015.
32. "Sushi Edging Pacific Bluefin Tuna Toward Extinction" DNews 19 February 2015.
33. Satran, Joe. "Pacific Bluefin Tuna Overfishing Has Led To 96 Percent Population Reduction, Study Says" The Huffington Post. TheHuffingtonPost.com. 19 February 2015
34. "European Eel Is Critically Endangered National Biodiversity Data Centre." European Eel Is Critically Endangered | National Biodiversity Data Centre. 19 February 2015.
35. Morris, Steve. "Eels in Crisis after 95% Decline in Last 25 Years" The Guardian 30 April 2009. 19 February 2015.
36. "Johnson's Seagrass (Halophila Johnsonii) - Office of Protected Resources - NOAA Fisheries" Johnson's Seagrass (Halophila Johnsonii) - Office of Protected Resources - NOAA Fisheries. 1 March 2013. 19 February 2015.
37. Zacharias, Mark. "Marine Environmental Protection Policy: International Efforts to Address Threats to Marine Biodiversity." Marine Policy: An Introduction to Governance and International Law of the Oceans. 1st ed. 2014. Print.
38. Hinrichsen, Don. The Atlas of Coasts & Oceans: Ecosystems, Threatened Resources, Marine Conservation. Chicago: U of Chicago, 2011.
39. "IUCN, the International Union for Conservation of Nature." IUCN. 12 February 2015.
40. Smith, David. "Ballast Water" MITSG CCR: Marine Bioinvasions. 1 January 2006. 12 February 2015.
41. Choksi, Sejal. "The Hostile Takeover of San Francisco Bay" 1 May 2009. 19 February 2015.
42. Martin, Glen. "The Great Invaders / A New Ecosystem Is Evolving in San Francisco Bay. We Have No Idea What It Is, or Where It's Going" SFGate. 5 February 2006. 19 February 2015.
43. "Foreign Species Invade San Francisco Bay" NPR, 11 May 2011. 12 February 2015.
44. Foster, A.M., P. Fuller, A. Benson, S. Constant, D. Raikow, J. Larson, and A. Fusaro. 2015. Corbicula fluminea USGS Nonindigenous Aquatic Species Database, Gainesville, FL. Revision Date: 26 June 2014

CHAPTER 8

Coral Reef

Chapter Contents

- **Introduction**
- **Formation**
- **Zones**
- **Locations**
- **Biology**
- **Darwin's Paradox**
- **Biodiversity**
- **Importance**
- **Threats**
- **Protection**
- **Restoration**

A coral reef is a sea structure made out of living organisms that form a wall together. Stony corals are usually the main organism found in a reef, but other types of skeletal material also join together to form the thick walls. It is also "constructed" by the thousands of fish, sponges, and sea urchins grazing around the area. As these organisms move about, they actually help the reef settle and gain a more stable structure. Waves are also partly responsible for the wall's size and format, as they affect the geology of the formation.

INTRODUCTION

Coral reefs are diverse underwater ecosystems held together by calcium carbonate structures secreted by corals. Coral reefs are built by colonies of tiny animals found in marine waters that contain few nutrients. Most coral reefs are built from stony corals, which in turn consist of polyps that cluster in groups. The polyps belong to a group of animals known as Cnidaria, which also includes sea anemones and jellyfish. Unlike sea anemones, corals secrete hard carbonate exoskeletons which support and protect the coral polyps. Most reefs grow best in warm, shallow, clear, sunny and agitated waters.

Often called «rainforests of the sea», shallow coral reefs form some of the most diverse ecosystems on Earth. They occupy less than 0.1% of the world's ocean surface, about half the area of France, yet they provide a home for at least 25% of all marine species, including fish, mollusks, worms, crustaceans, echinoderms, sponges, tunicates and other cnidarians. Paradoxically, coral reefs flourish even though they are surrounded by ocean waters that provide few nutrients. They are most commonly found at shallow depths in tropical waters, but deep water and cold water corals also exist on smaller scales in other areas.

Coral reefs deliver ecosystem services to tourism, fisheries and shoreline protection. The annual global economic value of coral reefs is estimated between US$29.8-375 billion. However, coral reefs are fragile ecosystems, partly because they are very sensitive to water temperature. They are under threat from climate change, oceanic acidification, blast fishing, cyanide fishing for aquarium fish, sunscreen use, overuse of reef resources, and harmful land-use practices, including urban and agricultural runoff and water pollution, which can harm reefs by encouraging excess algal growth.

FORMATION

Most of the coral reefs we can see today were formed after the last glacial period when melting ice caused the sea level to rise and flood the continental shelves. This means that most modern coral reefs are less than 10,000 years old. As communities established themselves on the shelves, the reefs grew upwards, pacing rising sea levels. Reefs that rose too slowly could become drowned reefs. They are covered by so much water that there was insufficient light. Coral reefs are found in the deep sea away from continental shelves, around oceanic islands and as atolls. The vast majority of these islands are volcanic in origin. The few exceptions have tectonic origins where plate movements have lifted the deep ocean floor on the surface.

In 1842 in his first monograph, *The Structure and Distribution of Coral Reefs*, Charles Darwin set out his theory of the formation of atoll reefs, an idea he conceived during the voyage of the *Beagle*. He theorized uplift and subsidence of the Earth's crust under the oceans formed the atolls. Darwin's theory sets out a sequence of three stages in atoll formation. It starts with a fringing reef forming around an extinct volcanic island as the island and ocean floor subsides. As the subsidence continues, the fringing reef becomes a barrier reef, and ultimately an atoll reef.

Darwin's theory starts with a volcanic island which becomes extinct

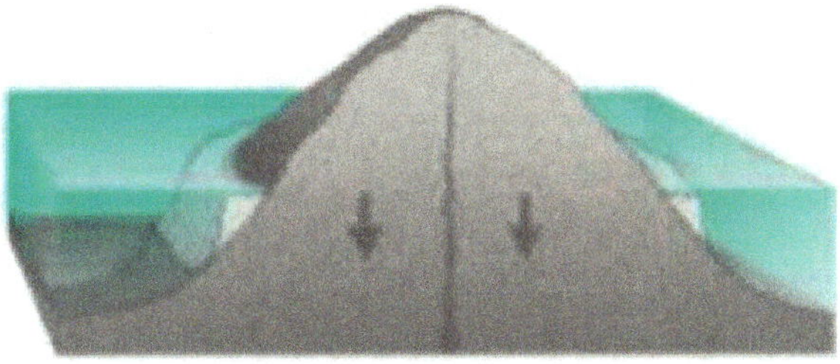

As the island and ocean floor subside, coral growth builds a fringing reef, often including a shallow lagoon between the land and the main reef

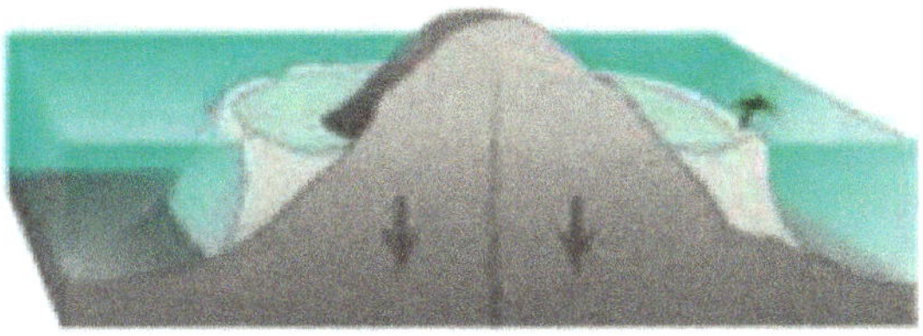

As the subsidence continues, the fringing reef becomes a larger barrier reef further from the shore with a bigger and deeper lagoon inside

Ultimately, the island sinks below the sea, and the barrier reef becomes an atoll enclosing an open lagoon

Darwin predicted that underneath each lagoon would be a bed rock base, the remains of the original volcano. Subsequent drilling proved this correct. Darwin's theory followed from his understanding that coral polyps thrive in the clean seas of the tropics where the water is agitated, but can only live within a limited depth range, starting just below low tide. Where the level of the underlying earth allows, the corals grow around the coast to form what he called fringing reefs, and can eventually grow out from the shore to become a barrier reef.

A fringing reef can take ten thousand years to form, and an atoll can take up to 30 million years

Where the bottom is rising, fringing reefs can grow around the coast, but coral raised above sea level dies and becomes white limestone. If the land subsides slowly, the fringing reefs keep pace by growing upwards on a base of older, dead coral, forming a barrier reef enclosing a lagoon between the reef and the land. A barrier reef can encircle an island, and once the island sinks below sea level a roughly circular atoll of growing coral continues to keep up with the sea level, forming a central lagoon. Barrier reefs and atolls do not usually form complete circles, but are broken in places by storms. Like sea level rise, a rapidly subsiding bottom can overwhelm coral growth, killing the coral polyps and the reef, due to what is called *coral drowning*. Corals that rely on zooxanthellae can *drown* when the water becomes too deep for their symbionts to adequately photosynthesize, due to decreased light exposure.

The two main variables determining the geomorphology, or shape, of coral reefs are the nature of the underlying substrate on which they rest, and the history of the change in sea level relative to that substrate.

The approximately 20,000-year-old Great Barrier Reef offers an example of how coral reefs formed on continental shelves. Sea level was then 120 m (390 ft) lower than in the 21st century. As sea level rose, the water and the corals encroached on what had been hills of the Australian coastal plain. By 13,000 years ago, sea level had risen to 60 m (200 ft) lower than at present, and many hills of the coastal plains had become continental islands. As the sea level rise continued, water topped most of the continental islands. The corals could then overgrow the hills, forming the present cays and reefs. Sea level on the Great Barrier Reef has not changed significantly in the last 6,000 years, and the age of the modern living reef structure is estimated to be between 6,000 and 8,000 years. Although the Great Barrier Reef formed along a continental shelf, and not around a volcanic island, Darwin's principles apply. Development stopped at the barrier reef stage, since Australia is not about to submerge. It formed the world's largest barrier reef, 300–1,000 m (980–3,280 ft) from shore, stretching for 2,000 km (1,200 mi).

Healthy tropical coral reefs grow horizontally from 1 to 3 cm (0.39 to 1.18 in) per year, and grow vertically anywhere from 1 to 25 cm (0.39 to 9.84 in) per year; however, they grow only at depths shallower than 150 m (490 ft) because of their need for sunlight, and cannot grow above sea level.

Materials

As the name implies, the bulk of coral reefs is made up of coral skeletons from mostly intact coral colonies. As other chemical elements present in corals become incorporated into the calcium carbonate deposits, aragonite is formed. However, shell fragments and the remains of calcareous algae such as the green-segmented genus *Halimeda* can add to the reef's ability to withstand damage from storms and other threats. Such mixtures are visible in structures such as Eniwetok Atoll.

Types

The three principal reef types are:

- **Fringing reef** – directly attached to a shore, or borders it with an intervening shallow channel or lagoon
- **Barrier reef** – reef separated from a mainland or island shore by a deep channel or lagoon
- **Atoll reef** – more or less circular or continuous barrier reef extends all the way around a lagoon without a central island

A small atoll in the Maldives

Inhabited cay in the Maldives

Other reef types or variants are:

- **Patch reef** – common, isolated, comparatively small reef outcrop, usually within a lagoon or embayment, often circular and surrounded by sand or seagrass
- **Apron reef** – short reef resembling a fringing reef, but more sloped; extending out and downward from a point or peninsular shore
- **Bank reef** – linear or semicircular shaped-outline, larger than a patch reef
- **Ribbon reef** – long, narrow, possibly winding reef, usually associated with an atoll lagoon
- **Table reef** – isolated reef, approaching an atoll type, but without a lagoon
- **Habili** – reef specific to the Red Sea; does not reach the surface near enough to cause visible surf; may be a hazard to ships (from the Arabic for "unborn")
- **Microatoll** – community of species of corals; vertical growth limited by average tidal height; growth morphologies offer a low-resolution record of patterns of sea level change; fossilized remains can be dated using radioactive carbon dating and have been used to reconstruct Holocene sea levels
- **Cays** – small, low-elevation, sandy islands formed on the surface of coral reefs from eroded material that piles up, forming an area above sea level; can be stabilized by plants to become habitable; occur in tropical environments throughout the Pacific, Atlantic and Indian Oceans (including the Caribbean and on the Great Barrier Reef and Belize Barrier Reef), where they provide habitable and agricultural land
- **Seamount** or **guyot** – formed when a coral reef on a volcanic island subsides; tops of seamounts are rounded and guyots are flat; flat tops of guyots, or *tablemounts*, are due to erosion by waves, winds, and atmospheric processes

ZONES

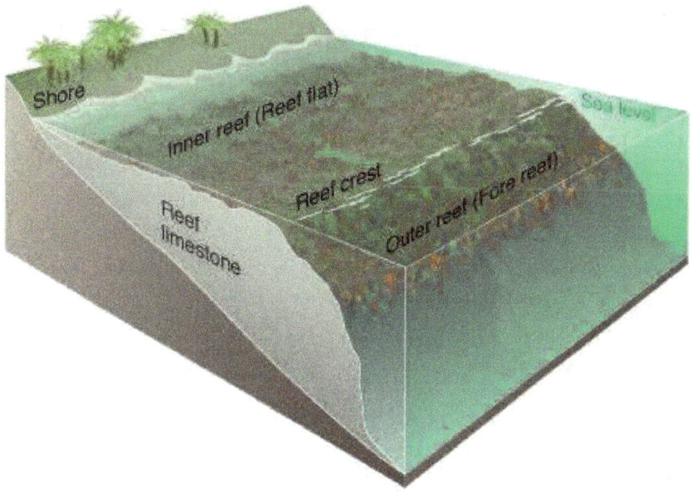

The three major zones of a coral reef: the fore reef, reef crest, and the back reef

Coral reef ecosystems contain distinct zones that represent different kinds of habitats. Usually, three major zones are recognized: the fore reef, reef crest, and the back reef (frequently referred to as the reef lagoon).

All three zones are physically and ecologically interconnected. Reef life and oceanic processes create opportunities for exchange of seawater, sediments, nutrients, and marine life among one another.

Thus, they are integrated components of the coral reef ecosystem, each playing a role in the support of the reefs' abundant and diverse fish assemblages.

Most coral reefs exist in shallow waters less than 50 m deep. Some inhabit tropical continental shelves where cool, nutrient rich upwelling does not occur, such as Great Barrier Reef. Others are found in the deep ocean surrounding islands or as atolls, such as in the Maldives. The reefs surrounding islands form when islands subside into the ocean, and atolls form when an island subsides below the surface of the sea.

Alternatively, Moyle and Cech distinguish six zones, though most reefs possess only some of the zones.

Water in the reef surface zone is often agitated. This diagram represents a reef on a continental shelf. The water waves at the left travel over the *off-reef floor* until they encounter the *reef slope* or *fore reef*. Then the waves pass over the shallow *reef crest*. When a wave enters shallow water it shoals, that is, it slows down and the wave height increases.

The reef surface is the shallowest part of the reef. It is subject to the surge and the rise and fall of tides. When waves pass over shallow areas, they shoal, as shown in the diagram at the right. This means the water is often agitated. These are the precise condition under which corals flourish. Shallowness means there is plenty of light for photosynthesis by the symbiotic zooxanthellae, and agitated water promotes the ability of coral to feed on plankton. However, other organisms must be able to withstand the robust conditions to flourish in this zone.

- **The off-reef floor** is the shallow sea floor surrounding a reef. This zone occurs by reefs on continental shelves. Reefs around tropical islands and atolls drop abruptly to great depths, and do not have a floor. Usually sandy, the floor often supports seagrass meadows which are important foraging areas for reef fish.
- **The reef drop-off** is, for its first 50 m, habitat for many reef fish who find shelter on the cliff face and plankton in the water nearby. The drop-off zone applies mainly to the reefs surrounding oceanic islands and atolls.
- **The reef face** is the zone above the reef floor or the reef drop-off. This zone is often the most diverse area of the reef. Coral and calcareous algae growths provide complex habitats and areas which offer protection, such as cracks and crevices. Invertebrates and epiphytic algae provide much of the food for other organisms.
- **The reef flat** is the sandy-bottomed flat, which can be behind the main reef, containing chunks of coral. This zone may border a lagoon and serve as a protective area, or it may lie between the reef and the shore, and in this case is a flat, rocky area. Fishes tend to prefer living in that flat, rocky area, compared to any other zone, when it is present.
- **The reef lagoon** is an entirely enclosed region, which creates an area less affected by wave action that often contains small reef patches.

However, the «topography of coral reefs is constantly changing. Each reef is made up of irregular patches of algae, sessile invertebrates, and bare rock and sand. The size, shape and relative abundance of these patches changes from year to year in response to the various factors that favor one type of patch over another. Growing coral, for example, produces constant change in the fine structure of reefs. On a larger scale, tropical storms may knock out large sections of reef and cause boulders on sandy areas to move."

LOCATIONS

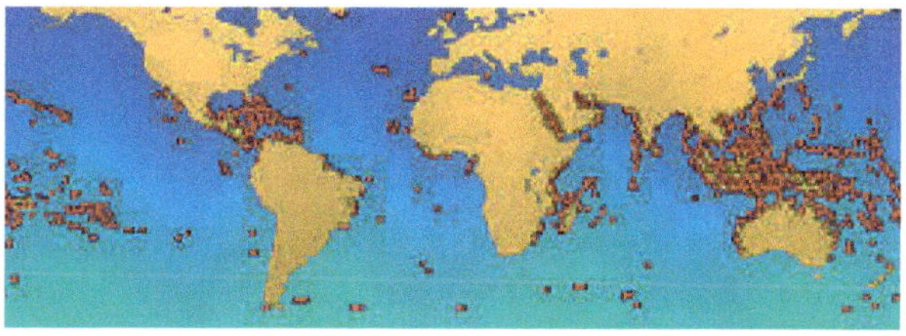

Locations of coral reefs

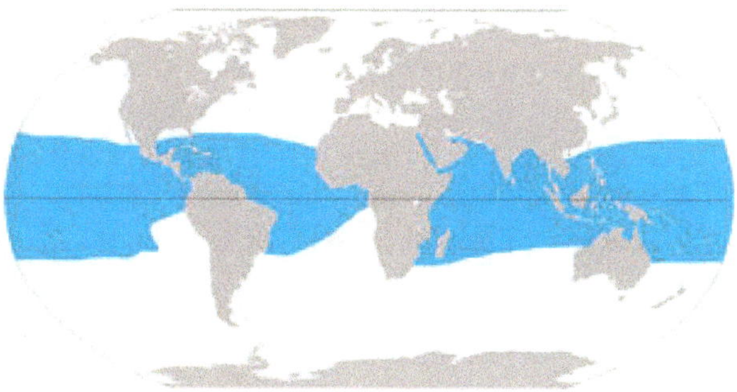

Boundary for 20 °C isotherms. Most corals live within this boundary. Note the cooler waters caused by upwelling on the southwest coast of Africa and off the coast of Peru

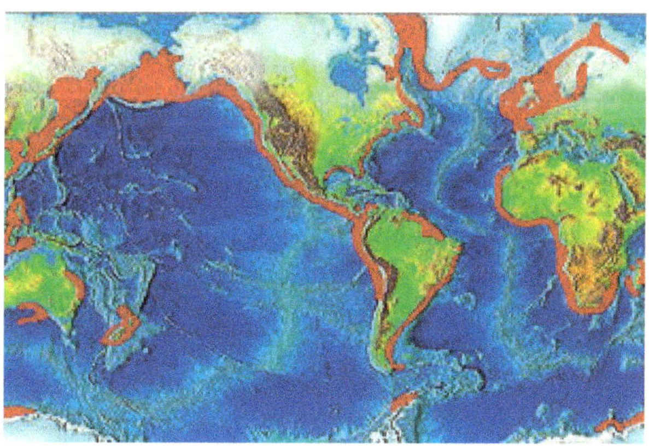

This map shows areas of upwelling in red. Coral reefs are not found in coastal areas where colder and nutrient-rich upwellings occur

Coral reefs are estimated to cover 284,300 km^2 (109,800 sq mi), just under 0.1% of the oceans› surface area. The Indo-Pacific region (including the Red Sea, Indian Ocean, Southeast Asia and the Pacific) account for 91.9% of this total. Southeast Asia accounts for 32.3% of that figure, while the Pacific including Australia accounts for 40.8%. Atlantic and Caribbean coral reefs account for 7.6%.

Although corals exist both in temperate and tropical waters, shallow-water reefs form only in a zone extending from approximately 30° N to 30° S of the equator. Tropical corals do not grow at depths of over 50 meters (160 ft). The optimum temperature for most coral reefs is 26–27 °C (79–81 °F), and few reefs exist in waters below 18 °C (64 °F). However, reefs in the Persian Gulf have adapted to temperatures of 13 °C (55 °F) in winter and 38 °C (100 °F) in summer. There are 37 species of scleractinian corals identified in such harsh environment around Larak Island.

Deep-water coral can exist at greater depths and colder temperatures at much higher latitudes, as far north as Norway. Although deep water corals can form reefs, very little is known about them.

Coral reefs are rare along the west coasts of the Americas and Africa, due primarily to upwelling and strong cold coastal currents that reduce water temperatures in these areas (respectively the Peru, Benguela and Canary streams). Corals are seldom found along the coastline of South Asia—from the eastern tip of India (Chennai) to the Bangladesh and Myanmar borders—as well as along the coasts of northeastern South America and Bangladesh, due to the freshwater release from the Amazon and Ganges Rivers respectively.

- The Great Barrier Reef—largest, comprising over 2,900 individual reefs and 900 islands stretching for over 2,600 kilometers (1,600 mi) off Queensland, Australia.
- The Mesoamerican Barrier Reef System—second largest, stretching 1,000 kilometers (620 mi) from Isla Contoy at the tip of the Yucatán Peninsula down to the Bay Islands of Honduras.
- The New Caledonia Barrier Reef—second longest double barrier reef, covering 1,500 kilometers (930 mi).
- The Andros, Bahamas Barrier Reef—third largest, following the east coast of Andros Island, Bahamas, between Andros and Nassau.
- The Red Sea—includes 6000-year-old fringing reefs located around a 2,000 km (1,240 mi) coastline.
- The Florida Reef Tract—largest continental US reef, extends from Soldier Key, located in Biscayne Bay, to the Dry Tortugas in the Gulf of Mexico.
- Pulley Ridge—deepest photosynthetic coral reef, Florida.
- Numerous reefs scattered over the Maldives.
- The Philippines coral reef area, the second largest in Southeast Asia, is estimated at 26,000 square kilometers and holds an extraordinary diversity of species. Scientists have identified 915 reef fish species and more than 400 scleractinian coral species, 12 of which are endemic.
- The Raja Ampat Islands in Indonesia`s West Papua province offer the highest known marine diversity.
- Bermuda is known for its northernmost coral reef system, located at 32.4° N and 64.8° W. The presence of coral reefs at this high latitude is due to the

proximity of the Gulf Stream. Bermuda has a fairly consistent diversity of coral species, representing a subset of those found in the greater Caribbean.

- The world›s northernmost individual coral reef so far discovered is located within a bay of Japan›s Tsushima Island in the Korea Strait.
- The world›s southernmost coral reef is at Lord Howe Island, in the Pacific Ocean off the east coast of Australia.

BIOLOGY

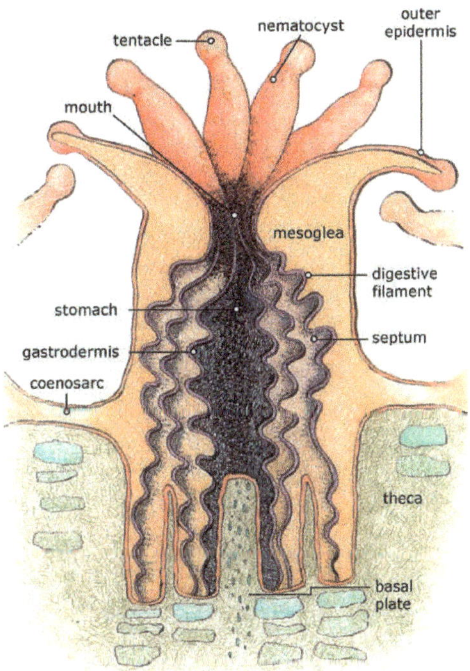

Anatomy of a coral polyp

Alive corals are colonies of small animals embedded in calcium carbonate shells. It is a mistake to think of coral as plants or rocks. Coral heads consist of accumulations of individual animals called polyps, arranged in diverse shapes. Polyps are usually tiny, but they can range in size from a pinhead to 12 inches (30 cm) across.

Reef-building or hermatypic corals live only in the photic zone (above 50 m), the depth to which sufficient sunlight penetrates the water, allowing photosynthesis to occur. Coral polyps do not photosynthesize, but have a symbiotic relationship with microscopic algae of the genus *Symbiodinium*, commonly referred to as zooxanthellae. These organisms live within the tissues of polyps and provide organic nutrients that nourish the polyp. Because of this relationship, coral reefs grow much faster in clear water, which admits more sunlight. Without their symbionts, coral growth would be too slow to form significant reef structures. Corals get up to 90% of their nutrients from their symbionts.

Reefs grow as polyps and other organisms deposit calcium carbonate, the basis of coral, as a skeletal structure beneath and around themselves, pushing the coral head's top upwards and outwards. Waves, grazing fish (such as parrotfish), sea urchins, sponges, and other forces and organisms act as bioeroders, breaking down coral skeletons into fragments that settle into spaces in the reef structure or form sandy bottoms in associated reef lagoons. Many other organisms living in the reef community contribute skeletal calcium carbonate in the same manner. Coralline algae are important contributors to reef structure in those parts of the reef subjected to the greatest forces by waves (such as the reef front facing the open ocean). These algae strengthen the reef structure by depositing limestone in sheets over the reef surface.

Typical shapes for coral species are wrinkled brains, cabbages, table tops, antlers, wire strands and pillars. These shapes can depend on the life history of the coral, like light exposure and wave action, and events such as breakages.

Table coral

Close up of polyps are arrayed on a coral, waving their tentacles. There can be thousands of polyps on a single coral branch

Corals reproduce both sexually and asexually. An individual polyp uses both reproductive modes within its lifetime. Corals reproduce sexually by either internal or external fertilization. The reproductive cells are found on the mesenteries, membranes that radiate inward from the layer of tissue that lines the stomach cavity. Some mature adult corals are hermaphroditic; others are exclusively male or female. A few species change sex as they grow.

Internally fertilized eggs develop in the polyp for a period ranging from days to weeks. Subsequent development produces a tiny larva, known as a planula. Externally fertilized eggs develop during synchronized spawning. Polyps release eggs and sperm into the water en masse, simultaneously. Eggs disperse over a large area. The timing of spawning depends on time of year, water temperature, and tidal and lunar cycles. Spawning is most successful when there is little variation between high and low tide. The less water movement, the better the chance for fertilization. Ideal timing occurs in the spring. Release of eggs or planula usually occurs at night, and is sometimes in phase with the lunar cycle (three to six days after a full moon). The period from release to settlement lasts only a few days, but some planulae can survive afloat for several weeks. They are vulnerable to predation and environmental conditions. The lucky few planulae which successfully attach to substrate next confront competition for food and space.

There are eight clades of Symbiodinium phylotypes. Most research has been completed on the Symbiodinium clades A–D. Each one of the eight contributes their own benefits as well as less compatible attributes to the survival of their coral hosts. Each photosynthetic organism has a specific level of sensitivity to photodamage of compounds needed for survival, such as proteins. Rates of regeneration and replication determine the organism›s ability to survive. Phylotype A is found more in the shallow regions of marine waters. It is able to produce mycosporine-like amino acids that are UV resistant, using a derivative of glycerin to absorb the UV radiation and allowing them to become more receptive to warmer water temperatures. In the event of UV or thermal damage, if and when repair occurs, it will increase the likelihood of survival of the host and symbiont. This leads to the idea that, evolutionarily, clade A is more UV resistant and thermally resistant than the other clades.

Clades B and C are found more frequently in the deeper water regions, which may explain the higher susceptibility to increased temperatures. Terrestrial plants that receive less sunlight because they are found in the undergrowth can be analogized to clades B, C, and D. Since clades B through D are found at deeper depths, they require an elevated light absorption rate to be able to synthesize as much energy. With elevated absorption rates at UV wavelengths, the deeper occurring phylotypes are more prone to coral bleaching versus the more shallow clades. Clade D has been observed to be high temperature-tolerant, and as a result it has a higher rate of survival than clades B and C.

Brain coral

Staghorn coral

Spiral wire coral

Pillar coral

Mushroom coral

Maze coral

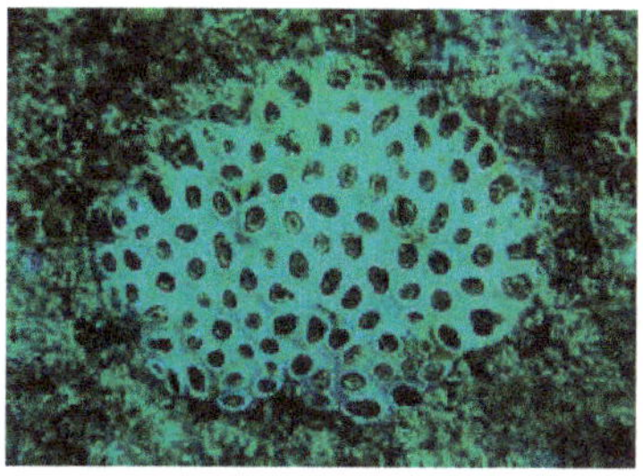

Black coral

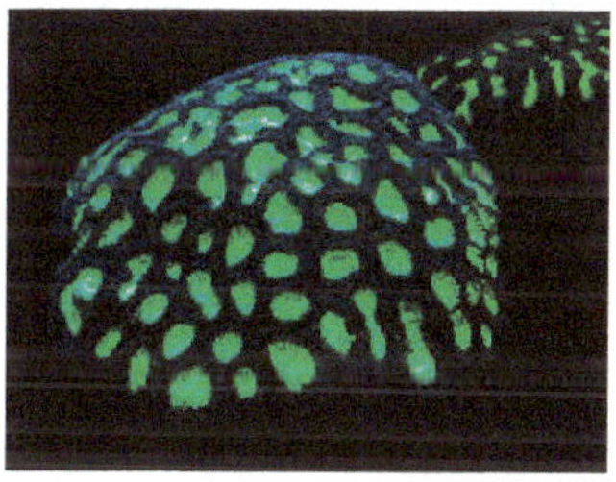

Fluorescent coral

DARWIN'S PARADOX

"Coral... seems to proliferate when ocean waters are warm, poor, clear and agitated, a fact which Darwin had already noted when he passed through Tahiti in 1842. This constitutes a fundamental paradox, shown quantitatively by the apparent impossibility of balancing input and output of the nutritive elements which control the coral polyp metabolism.

Recent oceanographic research has brought to light the reality of this paradox by confirming that the oligotrophy of the ocean euphotic zone persists right up to the swell-battered reef crest. When you approach the reef edges and atolls from the quasidesert of the open sea, the near absence of living matter suddenly becomes a plethora of life, without transition. So why is there something rather than nothing, and more precisely, where do the necessary nutrients for the functioning of this extraordinary coral reef machine come from?» — Francis Rougerie

In *The Structure and Distribution of Coral Reefs*, published in 1842, Darwin described how coral reefs were found in some areas of the tropical seas but not others, with no obvious cause. The largest and strongest corals grew in parts of the reef exposed to the most violent surf and corals were weakened or absent where loose sediment accumulated.

Tropical waters contain few nutrients yet a coral reef can flourish like an "oasis in the desert". This has given rise to the ecosystem conundrum, sometimes called "Darwin's paradox": "How can such high production flourish in such nutrient poor conditions?"

Coral reefs cover less than 0.1% of the surface of the world's ocean, about half the land area of France, yet they support over one-quarter of all marine species. This diversity results in complex food webs, with large predator fish eating smaller forage fish that eat yet smaller zooplankton and so on. However, all food webs eventually depend on plants, which are the primary producers. Coral reefs, primary productivity is very high, typically producing 5–10 grams of carbon per square meter per day ($gC \cdot m^{-2} \cdot day^{-1}$) biomass.

One reason for the unusual clarity of tropical waters is they are deficient in nutrients and drifting plankton. Further, the sun shines year-round in the tropics, warming the surface layer, making it less dense than subsurface layers. The warmer water is separated from deeper, cooler water by a stable thermocline, where the temperature makes a rapid change. This keeps the warm surface waters floating above the cooler deeper waters. In most parts of the ocean, there is little exchange between these layers. Organisms that die in aquatic environments generally sink to the bottom, where they decompose, which releases nutrients in the form of nitrogen (N), phosphorus (P) and potassium (K). These nutrients are necessary for plant growth, but in the tropics, they do not directly return to the surface.

Coral polyps

Plants form the base of the food chain, and need sunlight and nutrients to grow. In the ocean, these plants are mainly microscopic phytoplankton which drift in the water column. They need sunlight for photosynthesis, which powers carbon fixation, so they are found only relatively near the surface. But they also need nutrients. Phytoplankton rapidly use nutrients in the surface waters, and in the tropics, these nutrients are not usually replaced because of the thermocline.

Explanations

Around coral reefs, lagoons fill in with material eroded from the reef and the island. They become havens for marine life, providing protection from waves and storms.

Most importantly, reefs recycle nutrients, which happens much less in the open ocean. In coral reefs and lagoons, producers include phytoplankton, as well as seaweed and coralline algae, especially small types called turf algae, which pass nutrients to corals. The phytoplankton are eaten by fish and crustaceans, who also pass nutrients along the food web. Recycling ensures fewer nutrients are needed overall to support the community.

Coral reefs support many symbiotic relationships. In particular, zooxanthellae provide energy to coral in the form of glucose, glycerol, and amino acids. Zooxanthellae can provide up to 90% of a coral's energy requirements. In return, as an example of mutualism, the corals shelter the zooxanthellae, averaging one million for every cubic centimeter of coral, and provide a constant supply of the carbon dioxide they need for photosynthesis.

The color of corals depends on the combination of brown shades provided by their zooxanthellae and pigmented proteins (reds, blues, greens, etc.) produced by the corals themselves

Corals also absorb nutrients, including inorganic nitrogen and phosphorus, directly from water. Many corals extend their tentacles at night to catch zooplankton that brush them when the water is agitated. Zooplankton provide the polyp with nitrogen, and the polyp shares some of the nitrogen with the zooxanthellae, which also require this element. The varying pigments in different species of zooxanthellae give them an overall brown or golden-brown appearance, and give brown corals their colors. Other pigments such as reds, blues, greens, etc. come from colored proteins made by the coral animals. Coral which loses a large fraction of its zooxanthellae becomes white (or sometimes pastel shades in corals that are richly pigmented with their own colorful proteins) and is said to be bleached, a condition which, unless corrected, can kill the coral.

Sponges are another key: they live in crevices in the coral reefs. They are efficient filter feeders, and in the Red Sea they consume about 60% of the phytoplankton that drifts by. The sponges eventually excrete nutrients in a form the corals can use.

Most coral polyps are nocturnal feeders. Here, in the dark, polyps have extended their tentacles to feed on zooplankton

The roughness of coral surfaces is the key to coral survival in agitated waters. Normally, a boundary layer of still water surrounds a submerged object, which acts as a barrier. Waves breaking on the extremely rough edges of corals disrupt the boundary layer, allowing the corals access to passing nutrients. Turbulent water thereby promotes reef growth and branching. Without the nutritional gains brought by rough coral surfaces, even the most effective recycling would leave corals wanting in nutrients.

Studies have shown that deep nutrient-rich water entering coral reefs through isolated events may have significant effects on temperature and nutrient systems. This water movement disrupts the relatively stable thermocline that usually exists between warm shallow water to deeper colder water. Leichter et al. (2006) found that temperature regimes on coral reefs in the Bahamas and Florida were highly variable with temporal scales of minutes to seasons and spatial scales across depths.

Water can be moved through coral reefs in various ways, including current rings, surface waves, internal waves and tidal changes. Movement is generally created by tides and wind. As tides interact with varying bathymetry and wind mixes with surface water, internal waves are created. An internal wave is a gravity wave that moves along density stratification within the ocean. When a water parcel encounters a different density it will oscillate and create internal waves. While internal waves generally have a lower frequency than surface waves, they often form as a single wave that breaks into multiple waves as it hits a slope and moves upward. This vertical break up of internal waves causes significant diapycnal mixing and turbulence. Internal waves can act as nutrient pumps, bringing plankton and cool nutrient-rich water up to the surface.

The irregular structure characteristic of coral reef bathymetry may enhance mixing and produce pockets of cooler water and variable nutrient content. Arrival of cool, nutrient-rich water from depths due to internal waves and tidal bores has been linked to growth rates of suspension feeders and benthic algae as well as plankton and larval organisms. Leichter et al. proposed that Codium isthmocladum react to deep water nutrient sources due to their tissues having different concentrations of nutrients dependent upon depth. Wolanski and

Hamner noted aggregations of eggs, larval organisms and plankton on reefs in response to deep water intrusions. Similarly, as internal waves and bores move vertically, surface-dwelling larval organisms are carried toward the shore. This has significant biological importance to cascading effects of food chains in coral reef ecosystems and may provide yet another key to unlocking "Darwin's Paradox".

Cyanobacteria provide soluble nitrates for the reef via nitrogen fixation.

Coral reefs also often depend on surrounding habitats, such as seagrass meadows and mangrove forests, for nutrients. Seagrass and mangroves supply dead plants and animals which are rich in nitrogen and also serve to feed fish and animals from the reef by supplying wood and vegetation. Reefs, in turn, protect mangroves and seagrass from waves and produce sediment in which the mangroves and seagrass can root.

BIODIVERSITY

Tube sponges attracting cardinal fishes, glassfishes and wrasses

Over 4,000 species of fish inhabit coral reefs

Organisms can cover every square inch of a coral reef

Coral reefs form some of the world's most productive ecosystems, providing complex and varied marine habitats that support a wide range of other organisms. Fringing reefs just below low tide level have a mutually beneficial relationship with mangrove forests at high tide level and sea grass meadows in between: the reefs protect the mangroves and seagrass from strong currents and waves that would damage them or erode the sediments in which they are rooted, while the mangroves and sea grass protect the coral from large influxes of silt, fresh water and pollutants. This level of variety in the environment benefits many coral reef animals, which, for example, may feed in the sea grass and use the reefs for protection or breeding.

Reefs are home to a large variety of animals, including fish, seabirds, sponges, cnidarians (which includes some types of corals, and jellyfish), worms, crustaceans (including shrimp, cleaner shrimp, spiny lobsters and crabs), mollusks (including cephalopods), echinoderms (including starfish, sea urchins, and sea cucumbers), sea squirts, sea turtles and sea snakes. Aside from humans, mammals are rare on coral reefs, with visiting cetaceans such as dolphins being the main exception. A few of these varied species feed directly on corals, while others graze on algae on the reef. Reef biomass is positively related to species diversity. The same hideouts in a reef may be regularly inhabited by different species at different times of day. Nighttime predators such as cardinalfish and squirrelfish hide during the day, while damselfish, surgeonfish, triggerfish, wrasses and parrotfish hide from eels and sharks.

Algae

Reefs are chronically at risk of algal encroachment. Overfishing and excess nutrient supply from onshore can enable algae to outcompete and kill the coral. Increased nutrient levels can be a result of sewage or chemical fertilizer runoff from nearby coastal developments. Runoff can carry nitrogen and phosphorus which promote excess algae growth. Algae can sometimes out-compete the coral for space. The algae can then smother the coral by decreasing the oxygen supply available to the reef. Decreased oxygen levels can slow down coral's calcification rates weakening the coral and leaving it more susceptible to disease and degradation. In surveys done around largely uninhabited US Pacific islands, algae inhabit a large percentage of surveyed coral locations. The algal population consists of turf algae, coralline algae, and macro algae.

Sponges

Sponges are essential for the functioning of the coral reef's ecosystem. Algae and corals in coral reefs produce organic material. This is filtered through sponges which convert this organic material into small particles which in turn are absorbed by algae and corals.

Fish

Over 4,000 species of fish inhabit coral reefs. The reasons for this diversity remain unclear. Hypotheses include the "lottery", in which the first (lucky winner) recruit to a territory is typically able to defend it against latecomers, "competition", in which adults compete for territory, and less-competitive species must be able to survive in poorer habitat, and "predation", in which population size is a function of postsettlement piscivore mortality. Healthy reefs can produce up to 35 tons of fish per square kilometer each year, but damaged reefs produce much less.

Invertebrates

Sea urchins, Dotidae and sea slugs eat seaweed. Some species of sea urchins, such as *Diadema antillarum*, can play a pivotal part in preventing algae from overrunning reefs. Nudibranchia and sea anemones eat sponges.

A number of invertebrates, collectively called «cryptofauna,» inhabit the coral skeletal substrate itself, either boring into the skeletons (through the process of bioerosion) or living in pre-existing voids and crevices. Those animals boring into the rock include sponges, bivalve mollusks, and sipunculans. Those settling on the reef include many other species, particularly crustaceans and polychaete worms.

Seabirds

Coral reef systems provide important habitats for seabird species, some endangered. For example, Midway Atoll in Hawaii supports nearly three million seabirds, including two-thirds (1.5 million) of the global population of Laysan albatross, and one-third of the global population of black-footed albatross. Each seabird species has specific sites on the atoll where they nest. Altogether, 17 species of seabirds live on Midway. The short-tailed albatross is the rarest, with fewer than 2,200 surviving after excessive feather hunting in the late 19th century.

Other

Sea snakes feed exclusively on fish and their eggs. Marine birds, such as herons, gannets, pelicans and boobies, feed on reef fish. Some land-based reptiles intermittently associate with reefs, such as monitor lizards, the marine crocodile and semiaquatic snakes, such as *Laticauda colubrina*. Sea turtles, particularly hawksbill sea turtles, feed on sponges.

Schooling reef fish

Caribbean reef squid

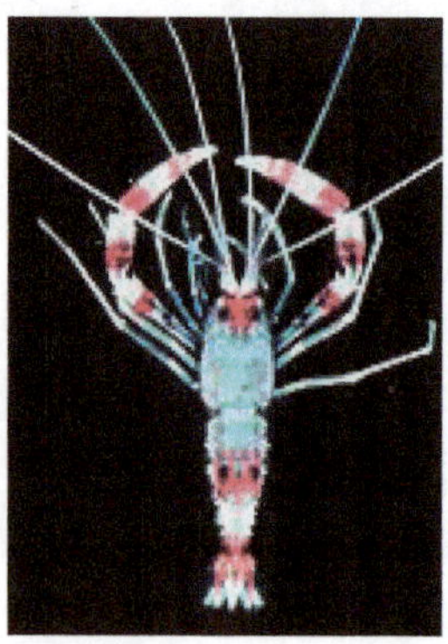

Banded coral shrimp

Whitetip reef shark

Green turtle

Giant clam

Soft coral, cup coral, sponges and ascidians

Banded sea krait

The shell of *Latiaxis wormaldi*, a coral snail

IMPORTANCE

Coral reefs deliver ecosystem services to tourism, fisheries and coastline protection. The global economic value of coral reefs has been estimated to be between US $29.8 billion and $375 billion per year. Coral reefs protect shorelines by absorbing wave energy, and many small islands would not exist without their reefs to protect them. According to the environmental group World Wide Fund for Nature, the economic cost over a 25-year period of destroying one kilometer of coral reef is somewhere between $137,000 and $1,200,000. About six million tons of fish are taken each year from coral reefs. Well-managed coral reefs have an annual yield of 15 tons of seafood on average per square kilometer. Southeast Asia›s coral reef fisheries alone yield about $2.4 billion annually from seafood.

To improve the management of coastal coral reefs, another environmental group, the World Resources Institute (WRI) developed and published tools for calculating the value of coral reef-related tourism, shoreline protection and fisheries, partnering with five Caribbean countries. As of April 2011, published working papers covered St. Lucia, Tobago, Belize, and the Dominican Republic, with a paper for Jamaica in preparation. The WRI was also "making sure that the study results support improved coastal policies and management planning". The Belize study estimated the value of reef and mangrove services at $395–559 million annually.

Bermuda›s coral reefs provide economic benefits to the Island worth on average $722 million per year, based on six key ecosystem services, according to Sarkis *et al* (2010).

THREATS

Island with fringing reef off Yap, Micronesia

Coral reefs are dying around the world. In particular, coral mining, agricultural and urban runoff, pollution (organic and inorganic), overfishing, blast fishing, disease, and the digging of canals and access into islands and bays are localized threats to coral ecosystems. Broader threats are sea temperature rise, sea level rise and pH changes from ocean acidification, all associated with greenhouse gas emissions. A 2014 study lists factors such as population explosion along the coast lines, overfishing, the pollution of coastal areas, global warming and invasive species among the main reasons that have put reefs in danger of extinction.

A study released in April 2013 has shown that air pollution can also stunt the growth of coral reefs; researchers from Australia, Panama and the UK used coral records (between 1880 and 2000) from the western Caribbean to show the threat of factors such as coal-burning coal and volcanic eruptions. Pollutants, such as Tributyltin, a biocide released into water from in anti-fouling paint can be toxic to corals.

In 2011, researchers suggested that «extant marine invertebrates face the same synergistic effects of multiple stressors» that occurred during the end-Permian extinction, and that genera «with poorly buffered respiratory physiology and calcareous shells», such as corals, were particularly vulnerable.

Rock coral on seamounts across the ocean are under fire from bottom trawling. Reportedly up to 50% of the catch is rock coral, and the practice transforms coral structures to rubble. With it taking years to regrow, these coral communities are disappearing faster than they can sustain themselves.

Another cause for the death of coral reefs is bioerosion. Various fishes graze corals, dead or alive and change the morphology of coral reefs making them more susceptible to other physical and chemical threats. It has been generally observed that only the algae growing on dead corals is eaten and the live ones are not. However, this act still destroys the top layer of coral substrate and makes it harder for the reefs to sustain.

In El Niño-year 2010, preliminary reports show global coral bleaching reached its worst level since another El Niño year, 1998, when 16% of the world›s reefs died as a result of increased water temperature. In Indonesia›s Aceh province, surveys showed some 80% of bleached corals died. Scientists do not yet understand the long-term impacts of coral bleaching, but they do know that bleaching leaves corals vulnerable to disease, stunts their growth, and affects their reproduction, while severe bleaching kills them. In July, Malaysia closed several dive sites where virtually all the corals were damaged by bleaching.

To find answers for these problems, researchers study the various factors that impact reefs. The list includes the ocean›s role as a carbon dioxide sink, atmospheric changes, ultraviolet light, ocean acidification, viruses, impacts of dust storms carrying agents to far-flung reefs, pollutants, algal blooms and others. Reefs are threatened well beyond coastal areas. Coral reefs with one type of zooxanthellae are more prone to bleaching than are reefs with another, more hardy, species.

General estimates show approximately 10% of the world›s coral reefs are dead. About 60% of the world›s reefs are at risk due to destructive, human-related activities. The threat to the health of reefs is particularly high in Southeast Asia, where 95% of reefs are at risk from local threats. By the 2030s, 90% of reefs are expected to be at risk from both human activities and climate change; by 2050, all coral reefs will be in danger.

Current research is showing that ecotourism in the Great Barrier Reef is contributing to coral disease, and that chemicals in sunscreens may contribute to the impact of viruses on zooxanthellae.

PROTECTION

A diversity of corals

Marine protected areas (MPAs) have become increasingly prominent for reef management. MPAs promote responsible fishery management and habitat protection. Much like national parks and wildlife refuges, and to varying degrees, MPAs restrict potentially damaging activities. MPAs encompass both social and biological objectives, including reef restoration, aesthetics, biodiversity, and economic benefits. However, there are very few MPAs that have actually made a substantial difference. Research in Indonesia, Philippines and Papua New Guinea shows that there is no significant difference between an MPA site and an unprotected site. Conflicts surrounding MPAs involve lack of participation, clashing views of the government and fisheries, effectiveness of the area, and funding. In some situations, as in the Phoenix Islands Protected Area, MPAs can also provide revenue, potentially equal to the income they would have generated without controls, as Kiribati did for its Phoenix Islands.

According to the *Caribbean Coral Reefs - Status Report 1970-2012* made by the IUCN. States that; stopping overfishing especially key fishes to coral reef like parrotfish, coastal zone management which reduce human pressure on reef, (for example restricting the coastal settlement, development and tourism in coastal reef) and controlling pollution specially sewage wastage, may not only reduce coral declining but also reverse it and may let to coral reef more adaptable to changes relates to climate and acidification. The report shows that healthier reef in the Caribbean are those with large population of parrotfish in countries which protect these key fishes and sea urchins, banning fish trap and Spearfishing creating "resilient reefs".

To help combat ocean acidification, some laws are in place to reduce greenhouse gases such as carbon dioxide. The Clean Water Act puts pressure on state government agencies to monitor and limit runoff of pollutants that can cause ocean acidification. Stormwater surge preventions are also in place, as well as coastal buffers between agricultural land and the coastline. This act also ensures that delicate watershed ecosystems are intact, such as wetlands. The Clean Water Act is funded by the federal government, and is monitored by various watershed groups. Many land use laws aim to reduce CO_2 emissions by limiting deforestation. Deforestation causes erosion, which releases a large amount of carbon stored in the soil, which then flows into the ocean, contributing to ocean acidification. Incentives are used to reduce miles traveled by vehicles, which reduces the carbon emissions into the atmosphere, thereby reducing the amount of dissolved CO_2 in the ocean. State and federal governments also control coastal erosion, which releases stored carbon in the soil into the ocean, increasing ocean acidification.High-end satellite technology is increasingly being employed to monitor coral reef conditions. Biosphere reserve, marine park, national monument and world heritage status can protect reefs. For example, Belize's barrier reef, Chagos archipelago, Sian Ka›an, the Galapagos islands, Great Barrier Reef, Henderson Island, Palau and Papahānaumokuākea Marine National Monument are world heritage sites.

In Australia, the Great Barrier Reef is protected by the Great Barrier Reef Marine Park Authority, and is the subject of much legislation, including a biodiversity action plan. They have compiled a Coral Reef Resilience Action Plan. This detailed action plan consists of numerous adaptive management strategies, including reducing our carbon footprint, which would ultimately reduce the amount of ocean acidification in the oceans surrounding the Great Barrier Reef. An extensive public awareness plan is also in place to provide education on the "rainforests of the sea" and how people can reduce carbon emissions, thereby reducing ocean acidification. Inhabitants of Ahus Island, Manus Province, Papua New Guinea, have followed a generations-old practice of restricting fishing in six areas of their reef lagoon. Their cultural traditions allow line fishing, but no net or spear fishing. The result is both the biomass and individual fish sizes are significantly larger than in places where fishing is unrestricted.

RESTORATION

Coral fragments growing on nontoxic concrete

Coral aquaculture, also known as coral farming or coral gardening, is showing promise as a potentially effective tool for restoring coral reefs, which have been declining around the world. The process bypasses the early growth stages of corals when they are most at risk of dying. Coral seeds are grown in nurseries, then replanted on the reef. Coral is farmed by coral farmers who live locally to the reefs and farm for reef conservation or for income.

Efforts to expand the size and number of coral reefs generally involve supplying substrate to allow more corals to find a home. Substrate materials include discarded vehicle tires, scuttled ships, subway cars, and formed concrete, such as reef balls. Reefs also grow unaided on marine structures such as oil rigs. In large restoration projects, propagated hermatypic coral on substrate can be secured with metal pins, superglue or milliput. Needle and thread can also attach A-hermatype coral to substrate.

A substrate for growing corals referred to as Biorock is produced by running low voltage electrical currents through seawater to crystallize dissolved minerals onto steel structures. The resultant white carbonate (aragonite) is the same mineral that makes up natural coral reefs. Corals rapidly colonize and grow at accelerated rates on these coated structures. The electrical currents also accelerate formation and growth of both chemical limestone rock and the skeletons of corals and other shell-bearing organisms. The vicinity of the anode and cathode provides a high-pH environment which inhibits the growth of competitive filamentous and fleshy algae. The increased growth rates fully depend on the accretion activity.

During accretion, the settled corals display an increased growth rate, size and density, but after the process is complete, growth rate and density return to levels comparable to natural growth, and are about the same size or slightly smaller.

One case study with coral reef restoration was conducted on the island of Oahu in Hawaii. The University of Hawaii has come up with a Coral Reef Assessment and Monitoring Program to help relocate and restore coral reefs in Hawaii. A boat channel on the island of Oahu to the Hawaii Institute of Marine Biology was overcrowded with coral reefs. Also, many areas of coral reef patches in the channel had been damaged from past dredging in the channel. Dredging covers the existing corals with sand, and their larvae cannot build and thrive on sand; they can only build on to existing reefs. Because of this, the University of Hawaii decided to relocate some of the coral reef to a different transplant site. They transplanted them with the help of the United States Army divers, to a relocation site relatively close to the channel. They observed very little, if any, damage occurred to any of the colonies while they were being transported, and no mortality of coral reefs has been observed on the new transplant site, but they will be continuing to monitor the new transplant site to see how potential environmental impacts (i.e. ocean acidification) will harm the overall reef mortality rate. While trying to attach the coral to the new transplant site, they found the coral placed on hard rock is growing considerably well, and coral was even growing on the wires that attached the transplant corals to the transplant site. This gives new hope to future research on coral reef transplant sites. As a result of this coral restoration project, no environmental effects were seen from the transplantation process, no recreational activities were decreased, and no scenic areas were affected by the project. This is a great example that coral transplantation and restoration can work and thrive under the right conditions, which means there may be hope for other damaged coral reefs.

Another possibility for coral restoration is gene therapy. Through infecting coral with genetically modified bacteria, it may be possible to grow corals that are more resistant to climate change and other threats.

Reefs in the Past

Ancient coral reefs

Throughout Earth history, from a few thousand years after hard skeletons were developed by marine organisms, there were almost always reefs. The times of maximum development were in the Middle Cambrian (513–501 Ma), Devonian (416–359 Ma) and Carboniferous (359–299 Ma), owing to order Rugosa extinct corals, and Late Cretaceous (100–66 Ma) and all Neogene (23 Ma–present), owing to order Scleractinia corals.

Not all reefs in the past were formed by corals: those in the Early Cambrian (542–513 Ma) resulted from calcareous algae and archaeocyathids (small animals with conical shape, probably related to sponges) and in the Late Cretaceous (100–66 Ma), when there also existed reefs formed by a group of bivalves called rudists; one of the valves formed the main conical structure and the other, much smaller valve acted as a cap.

Measurements of the oxygen isotopic composition of the aragonitic skeleton of coral reefs, such as Porites, can indicate changes in the sea surface temperature and sea surface salinity conditions of the ocean during the growth of the coral. This technique is often used by climate scientists to infer the paleoclimate of a region.

REFERENCES

1. Spalding MD, Grenfell AM (1997). "New estimates of global and regional coral reef areas". Coral Reefs. 16 (4): 225–230. doi:10.1007/s003380050078.
2. Spalding, Mark, Corinna Ravilious, and Edmund Green (2001). World Atlas of Coral Reefs. Berkeley, CA: University of California Press and UNEP/WCMC ISBN 0520232550.
3. Mulhall M (Spring 2009) Saving rainforests of the sea: An analysis of international efforts to conserve coral reefs Duke Environmental Law and Policy Forum 19:321–351.
4. Where are Corals Found? NOAA. Reviewed: 13 May 2011. Retrieved: 24 March 2015.
5. Hoover, John (November 2007). Hawaii's Sea Creatures. Mutual. ISBN 1-56647-220-2.
6. Cesar, H.J.S; Burke, L.; Pet-Soede, L. (2003). The Economics of Worldwide Coral Reef Degradation (PDF). The Netherlands: Cesar Environmental Economics Consulting. p. 4. Retrieved 21 September 2013.
7. Costanza, Robert; Ralph d'Arge; Rudolf de Groot; Stephen Farber; Monica Grasso; Bruce Hannon; Karin Limburg; Shahid Naeem; Robert V. O'Neill; Jose Paruelo; Robert G. Raskin; Paul Sutton; Marjan van den Belt (15 May 1997). "The value of the world's ecosystem services and natural capital". Nature. 387 (6630): 253–260. Bibcode:1997Natur.387..253C. doi:10.1038/387253a0.
8. Danovaro, Roberto; Bongiorni, Lucia; Corinaldesi, Cinzia; Giovannelli, Donato; Damiani, Elisabetta; Astolfi, Paola; Greci, Lucedio; Pusceddu, Antonio (April 2008). "Sunscreens Cause Coral Bleaching by Promoting Viral Infections". Environmental Health Perspectives. 116 (4): 441–447. doi:10.1289/ehp.10966. PMC 2291018Freely accessible. PMID 18414624.
9. "Corals reveal impact of land use". ARC Centre of Excellence for Coral Reef Studies. Retrieved September 21, 2013.
10. Minato, Charissa (July 1, 2002). "Urban runoff and coastal water quality being researched for effects on coral reefs" (PDF). Retrieved December 2010. Check date values in: |access-date= (help)
11. "Coastal Watershed Factsheets – Coral Reefs and Your Coastal Watershed". Environmental Protection Agency Office of Water. July 1998. Retrieved December 2010. Check date values in: |access-date= (help)
12. Kleypas, Joanie (September 21, 2010). "Coral reef". The Encyclopedia of Earth.

Retrieved April 4, 2011.

13. Darwin, Charles (1842). "The Structure and Distribution of Coral Reefs. Being the first part of the geology of the voyage of the Beagle, under the command of Capt. Fitzroy, R.N. during the years 1832 to 1836". London: Smith Elder and Co.
14. Chancellor, Gordon (2008). "Introduction to Coral reefs". Darwin Online. Retrieved January 20, 2009.
15. Animation of coral atoll formation NOAA Ocean Education Service. Retrieved January 9, 2010.
16. Webster, Jody M.; Braga, Juan Carlos; Clague, David A.; Gallup, Christina; Hein, James R.; Potts, Donald C.; Renema, Willem; Riding, Robert; Riker-Coleman, Kristin; Silver, Eli; Wallace, Laura M. (1 March 2009). "Coral reef evolution on rapidly subsiding margins". Global and Planetary Change. 66 (1–2): 129–148. Bibcode:2009GPC....66..129W. doi:10.1016/j.gloplacha.2008.07.010.
17. Webster, Jody M.; Clague, David A.; Riker-Coleman, Kristin; Gallup, Christina; Braga, Juan C.; Potts, Donald; Moore, James G.; Winterer, Edward L.; Paull, Charles K. (1 January 2004). "Drowning of the −150 m reef off Hawaii: A casualty of global meltwater pulse 1A?". Geology. 32 (3): 249. Bibcode:2004Geo....32..249W. doi:10.1130/G20170.1.
18. Great Barrier Reef Marine Park Authority (2006). "A "big picture" view of the Great Barrier Reef" (PDF). Reef Facts for Tour Guides. Retrieved June 18, 2007.
19. Tobin, Barry (2003) . "How the Great Barrier Reef was formed". Australian Institute of Marine Science. Retrieved November 22, 2006.
20. CRC Reef Research Centre Ltd. "What is the Great Barrier Reef?". Archived from the original on August 22, 2006. Retrieved May 28, 2006.
21. Four Types of Coral Reef Microdocs, Stanford Education. Retrieved January 10, 2010.
22. MSN Encarta (2006). Great Barrier Reef. Archived from the original on October 31, 2009. Retrieved December 11, 2006.
23. Murphy, Richard C. (2002). Coral Reefs: Cities Under The Seas. The Darwin Press, Inc. ISBN 0-87850-138-X.
24. Smithers, S.G.; Woodroffe, C.D. (2000). "Microatolls as sea-level indicators on a mid-ocean atoll". Marine Geology. 168 (1–4): 61–78. doi:10.1016/S0025-3227(00)00043-8.
25. Moyle, Peter B.; Joseph J. Cech (2004). Fishes : an introduction to ichthyology (Fifth ed.). Upper Saddle River, N.J.: Pearson/Prentice Hall. p. 556. ISBN 978-0-13-100847-2.
26. Connell, Joseph H. (March 24, 1978). "Diversity in Tropical Rain Forests and Coral Reefs". Science. 199 (4335): 1302–1310. Bibcode:1978Sci...199.1302C. doi:10.1126/science.199.4335.1302. PMID 17840770.
27. UNEP (2001) UNEP-WCMC World Atlas of Coral Reefs Coral Reef Unit
28. Achituv, Y. and Dubinsky, Z. 1990. Evolution and Zoogeography of Coral Reefs

Ecosystems of the World. Vol. 25:1–8.

29. Wells, Sue; Hanna, Nick (1992). Greenpeace Book of Coral Reefs. Sterling Publishing Company. ISBN 0-8069-8795-2.
30. Vajed Samiei, J.; Dab K.; Ghezellou P.; Shirvani A. (2013). "Some Scleractinian Corals (Class: Anthozoa) of Larak Island, Persian Gulf". Zootaxa. 3636 (1): 101–143. doi:10.11646/zootaxa.3636.1.5.
31. Gunnerus, Johan Ernst (1768). Om Nogle Norske Coraller.
32. Nybakken, James. 1997. Marine Biology: An Ecological Approach. 4th ed. Menlo Park, CA: Addison Wesley.
33. NOAA CoRIS – Regional Portal – Florida. Coris.noaa.gov (August 16, 2012). Retrieved on March 3, 2013.
34. NGM.nationalgeographic.com, Ultra Marine: In far eastern Indonesia, the Raja Ampat islands embrace a phenomenal coral wilderness, by David Doubilet, National Geographic, September 2007
35. Living Reefs Foundation. Retrieved on May 28, 2015.
36. LiveScience. Retrieved on April 14, 2016.
37. Sherman, C.D.H. "The Importance of Fine-scale Environmental Heterogeneity in Determining Levels of Genotypic Diversity and Local Adaption." University of Wollongong Ph.D. Thesis. 2006. Accessed June 7, 2009.
38. Marshall, Paul; Schuttenberg, Heidi (2006). A Reef Manager's Guide to Coral Bleaching. Townsville, Australia: Great Barrier Reef Marine Park Authority. ISBN 1-876945-40-0.
39. Stacy, J., Marion, G., McCulloch, M. and Hoegh-Guldberg, O. "changes to Mackay Whitsunday water quality and connectivity between terrestrial, mangrove and coral reef ecosystems: Clues from coral proxies and remote sensing records -Synthesis of research from an ARC Linkage Grant (2004–2007)." University of Queensland – Centre for Marine Studies. May 2007. Accessed June 7, 2009.
40. Nothdurft, L.D. "Microstructure and early diagenesis of recent reef building scleractinian corals, Heron Reef, Great Barrier Reef: Implications for palaeoclimate analysis." Queensland University of Technology Ph.D. Thesis. 2007. Accessed June 7, 2009.
41. Wilson, R.A. "The Biological Notion of Individual."Stanford Encyclopedia of Philosophy. August 9, 2007. Accessed June 7, 2009.
42. Jennings S, Kaiser MJ and Reynolds JD (2001) Marine fisheries ecology, Wiley-Blackwell, pp. 291–293. ISBN 978-0-632-05098-7.
43. Chappell, John (17 July 1980). "Coral morphology, diversity and reef growth". Nature. 286 (5770): 249–252. Bibcode:1980Natur.286..249C. doi:10.1038/286249a0.
44. Jackson, Jeremy B. C. (1 July 1991). "Adaptation and Diversity of Reef Corals". BioScience. 41 (7): 475–482. doi:10.2307/1311805. JSTOR 1311805.
45. Reynolds, J., B. Bruns, W. Fitt, G. Schmidt. 2008. Enhanced photoprotection

pathways in symbiotic dinoflagellates of shallow-water corals and other cnidarians. 13674–13678, doi:10.1073/pnas.0805187105

46. Fluorescent coral National Geographic.
47. Rougerier, F (1998). "The functioning of coral reefs and atolls: from paradox to paradigm". In Jost, Christian. The French-Speaking Pacific: Population, Environment and Development Issues (PDF). Boombana Publications. ISBN 978-1-876542-02-3.
48. Darwin, Charles R. (1842). The Structure and Distribution of Coral Reefs. Being the first part of the geology of the voyage of the Beagle, under the command of Capt. Fitzroy, R.N. during the years 1832 to 1836. London: Smith Elder and Co. pp. 61–71.
49. Crossland CJ (1983) "Dissolved nutrients in coral reef waters In DJ Barnes (Ed) Perspectives on coral reefs, pages 56–68, Australian Institute of Marine Science. ISBN 9780642895851.
50. Odum EP (1971) Fundamentals of ecology Saunders.
51. Sammarco, PW; Risk, MJ; Schwarcz, HP; Heikoop, JM (1999). "Cross-continental shelf trends in coral δ15N on the Great Barrier Reef: further consideration of the reef nutrient paradox" (PDF). Mar Ecol Prog Ser. 180: 131–138. doi:10.3354/meps180131.
52. Rougerie, F; Wauthy, B (1993). "The endo-upwelling concept: from geothermal convection to reef construction" (PDF). Coral Reefs. 12: 19–30. Bibcode:1993CorRe..12...19R. doi:10.1007/bf00303781.
53. De Goeij, Jasper M (2009) "Element cycling on tropical coral reefs: the cryptic carbon shunt revealed" PhD thesis, page 13. University of Groningen.
54. Sorokin, Yuri I. (1993). Coral Reef Ecology. Germany: Springer-Verlag, Berlin Heidelberg. ISBN 978-0-387-56427-2.
55. Hatcher, Bruce Gordon (1 May 1988). "Coral reef primary productivity: A beggar's banquet". Trends in Ecology & Evolution. 3 (5): 106–111. doi:10.1016/0169-5347(88)90117-6.
56. Castro, Peter and Huber, Michael (2000) Marine Biology. 3rd ed. Boston: McGraw-Hill.
57. Zooxanthellae… What's That?. Oceanservice.noaa.gov (March 25, 2008). Retrieved on November 1, 2011.
58. Roach, John (November 7, 2001). "Rich Coral Reefs in Nutrient-Poor Water: Paradox Explained?". National Geographic News. Retrieved April 5, 2011.
59. Nowak (21 September 2002). "Corals play rough over Darwin's paradox". New Scientist (2361). |first1= missing |last1= in Authors list (help)
60. Leichter, J.; Wing S.; Miller S.; Denny M. (1996). "Pulsed delivery of subthermocline water to Conch Reef (Florida Keys) by internal tidal bores". Limnology and Oceanography. 41 (7): 1490–1501. doi:10.4319/lo.1996.41.7.1490.
61. Wolanski, E.; Pickard, G. L. (1983). "Upwelling by internal tides and kelvin

waves at the continental shelf break on the Great Barrier Reef". Marine and Freshwater Research. 34: 65. doi:10.1071/MF9830065.

62. Leichter, J.; Helmuth B.; Fischer A. (2006). "Variation beneath the surface: Quantifying complex thermal environments on coral reefs in the Caribbean, Bahamas and Florida". Journal of Marine Research. 64 (4): 563–588. doi:10.1357/002224006778715711.
63. Ezer, T.; Heyman W.; Houser C.; Kjerfve B. (2011). "Modeling and observations of high-frequency flow variability and internal waves at a Caribbean reef spawning aggregation site". Ocean Dynamics. 61 (5): 581–598. Bibcode:2011OcDyn..61..581E. doi:10.1007/s10236-010-0367-2.
64. Fratantoni, D.; Richardson P. (2006). "The Evolution and Demise of North Brazil Current Rings". Journal of Physical Oceanography. 36 (7): 1241–1249. Bibcode:2006JPO....36.1241F. doi:10.1175/JPO2907.1.
65. Leichter, J.; Shellenbarger G.; Genovese S.; Wing S. (1998). "Breaking internal waves on a Florida (USA) coral reef: a plankton pump at work?". Marine Ecology Progress Series. 166: 83–97. doi:10.3354/meps166083.
66. Talley, L. (2011). Descriptive Physical Oceanography: An Introduction. Oxford UK: Elsevier Inc. ISBN 0750645520.
67. Helfrich, K. (1992). "Internal solitary wave breaking and run-up on a uniform slope". Journal of Fluid Mechanics. 243: 133–154. Bibcode:1992JFM...243..133H. doi:10.1017/S0022112092002660.
68. Gregg, M. (1989). "Scaling turbulent dissipation in the thermocline". Journal of Geophysical Research. 9686–9698. 94: 9686. Bibcode:1989JGR....94.9686G. doi:10.1029/JC094iC07p09686.
69. Taylor, J. (1992). "The energetics of breaking events in a resonantly forced internal wave field". Journal of Fluid Mechanics. 239: 309– 340. Bibcode:1992JFM...239..309T. doi:10.1017/S0022112092004427.
70. Andrews, J.; Gentien P. (1982). "Upwelling as a source of nutrients for the Great Barrier Reef ecosystems: A solution to Darwin's question?". Marine Ecology Progress Series. 8: 257–269. doi:10.3354/meps008257.
71. Sandstrom, H.; Elliott J. (1984). "Internal tide and solitons on the Scotian shelf: A nutrient pump at work". Journal of Geophysical Research. 89: 6415–6426. Bibcode:1984JGR....89.6415S. doi:10.1029/JC089iC04p06415.
72. Wolanski, E.; Hamner W. (1988). "Topographically controlled fronts in the ocean and their biological significance". Science. 241 (4862): 177–181. Bibcode:1988Sci...241..177W. doi:10.1126/science.241.4862.177. PMID 17841048.
73. Rougerie, F.; Fagerstrom J.; Andrie C. (1992). "Geothermal endo-upwelling: A solution to the reef nutrient paradox?". Continental Shelf Research. 12 (7–8): 785–798. Bibcode:1992CSR....12..785R. doi:10.1016/0278-4343(92)90044-K.
74. Wolanski, E.; Delesalle B. (1993). "Upwelling by internal waves, Tahiti, French Polynesia". Continental Shelf Research. 15 (2–3): 357–368.

Bibcode:1995CSR....15..357W. doi:10.1016/0278-4343(93)E0004-R.

75. Szmant, A. M.; Forrester, A. (1996). "Water column and sediment nitrogen and phosphorus distribution patterns in the Florida Keys, USA". Coral Reefs. 15: 21–41. Bibcode:1996CorRe..15...21S. doi:10.1007/BF01626075.

76. Furnas, M. J.; Mitchell, A. W. (1996). "Nutrient inputs into the central Great Barrier Reef (Australia) from subsurface intrusions of Coral Sea waters: A two-dimensional displacement model". Continental Shelf Research. 16 (9): 1127–1148. Bibcode:1996CSR....16.1127F. doi:10.1016/0278-4343(95)00060-7.

77. Leichter, J.; Miller S. (1999). "Predicting high-frequency upwelling: Spatial and temporal patterns of temperature anomalies on a Florida coral reef.". Continental Shelf Research. 19: 911–928. doi:10.1016/s0278-4343(99)00004-7.

78. Leichter, J.; Stewart H.; Miller S. (2003). "Episodic nutrient transport to Florida coral reefs". Limnology and Oceanography. 48 (4): 1394–1407. doi:10.4319/lo.2003.48.4.1394.

79. Leichter, J.; Deane G.; Stokes M. (2005). "Spatial and Temporal Variability of Internal Wave Forcing on a Coral Reef". Journal of Physical Oceanography. 35 (11): 1945–1962. Bibcode:2005JPO....35.1945L. doi:10.1175/JPO2808.1.

80. Smith, J.; Smith C.; Vroom P.; Beach K.; Miller S. (2004). "Nutrient and growth dynamics of Halimeda tuna on Conch Reef, Florida Keys: Possible influence of internal tides on nutrient status and physiology". Limnology and Oceanography. 49 (6): 1923–1936. doi:10.4319/lo.2004.49.6.1923.

81. Pineda, J. (1994). "Internal tidal bores in the nearshore: Warm-water fronts, seaward gravity currents and the onshore transport of neustonic larvae". Journal of Marine Research. 52 (3): 427–458. doi:10.1357/0022240943077046.

82. Wilson, E (2004). "Coral's Symbiotic Bacteria Fluoresce, Fix Nitrogen". Chemical and Engineering News. 82 (33): 7.

83. Barnes, R.S.K.; Mann, K.H. (1991). Fundamentals of Aquatic Ecology. Blackwell Publishing. pp. 217–227. ISBN 0-632-02983-8.

84. Fuchs. T (2013). "Effects of Coral Reef Complexity on Invertebrate Biodiversity". Immediate Science Ecology Publishing: 1–10.

85. Hatcher, B.G. Johannes, R.E.; Robertson, A.J. (1989). "Conservation of Shallow-water Marine Ecosystems". Oceanography and Marine Biology: An Annual Review. 27. Routledge. p. 320. ISBN 0-08-037718-1.

86. "World's Reef Fishes Tussling With Human Overpopulation". ScienceDaily. April 5, 2011.

87. "Coral Reef Biology". NOAA. Retrieved April 6, 2011.

88. Glynn, P.W. (1990). Dubinsky, Z., ed. Ecosystems of the World v. 25-Coral Reefs. New York, NY: Elsevier Science. ISBN 978-0-444-87392-7.

89. "THE EFFECTS OF TERRESTRIAL RUNOFF OF SEDIMENTS, NUTRIENTS AND OTHER POLLUTANTS ON CORAL REEFS" (PDF). Retrieved 2015-12-05.

90. Vroom, Peter S.; Page, Kimberly N.; Kenyon, Jean C.; Brainard, Russell

E. (2006). "Algae-Dominated Reefs". American Scientist. 94 (5): 430–437. doi:10.1511/2006.61.1004.

91. Kaplan, Matt (2009). "How the sponge stays slim". Nature. doi:10.1038/news.2009.1088.
92. Buchheim, Jason. "Coral Reef Fish Ecology". marinebiology.org. Retrieved April 5, 2011.
93. McClellan, Kate; Bruno, John (2008). "Coral degradation through destructive fishing practices". Encyclopedia of Earth. Retrieved October 25, 2008.
94. Osborne, Patrick L. (2000). Tropical Ecosystem and Ecological Concepts. Cambridge: Cambridge University Press. p. 464. ISBN 0-521-64523-9.
95. The.honoluluadvertiser.com. The.honoluluadvertiser.com (January 17, 2005). Retrieved on November 1, 2011.
96. "U.S. Fish & Wildlife Service – Birds of Midway Atoll". Retrieved August 19, 2009.
97. Heatwole, Harold (1999). Sea snakes (2. ed.). Malabar, Fla: Krieger. ISBN 1-57524-116-1.
98. Li, Min; Fry, B.G.; Kini, R. Manjunatha (1 January 2005). "Eggs-Only Diet: Its Implications for the Toxin Profile Changes and Ecology of the Marbled Sea Snake (Aipysurus eydouxii)". Journal of Molecular Evolution. 60 (1): 81–89. doi:10.1007/s00239-004-0138-0. PMID 15696370.
99. Voris, Harold K. (1 January 1966). "Fish Eggs as the Apparent Sole Food Item for a Genus of Sea Snake, Emydocephalus (Krefft)". Ecology. 47 (1): 152. doi:10.2307/1935755.
100. McClenachan, Loren; Jackson, Jeremy BC; Newman, Marah JH (1 August 2006). "Conservation implications of historic sea turtle nesting beach loss". Frontiers in Ecology and the Environment. 4 (6): 290–296. doi:10.1890/1540-9295(2006)4[290:CIOHST]2.0.CO;2.
101. Lutz, Peter L.; Musick, John A. (1996). The biology of sea turtles. Boca Raton, Fla: CRC Press. ISBN 0849384222.
102. Meylan, Anne (January 22, 1988). "Spongivory in Hawksbill Turtles: A Diet of Glass". Science. 239 (4838): 393–395. Bibcode:1988Sci...239..393M. doi:10.1126/science.239.4838.393. PMID 17836872.
103. "The Importance of Coral to People". World Wildlife Fund. Retrieved April 7, 2011.
104. "Coastal Capital: Economic Valuation of Coastal Ecosystems in the Caribbean". World Resources Institute.
105. Cooper, Emily; Burke, Lauretta; Bood, Nadia (2008). "Coastal Capital: Belize: The Economic Contribution of Belize's Coral Reefs and Mangroves" (PDF). Retrieved April 6, 2011.
106. Sarkis, Samia; van Beukering, Pieter J.H.; McKenzie, Emily (2010). "Total Economic Value of Bermuda's Coral Reefs. Valuation of ecosystem Services" (PDF). Retrieved May 29, 2015.

107. "Coral reefs around the world". Guardian.co.uk. September 2, 2009.

108. "Caribbean coral reefs may disappear within 20 years: Report". IANS. news. biharprabha.com. Retrieved 3 July 2014.

109. Liz Minchin (8 April 2013). "Air pollution casts a cloud over coral reef growth". The Conversation. The Conversation Media Group. Retrieved 9 April 2013.

110. Clapham ME and Payne (2011). "Acidification, anoxia, and extinction: A multiple logistic regression analysis of extinction selectivity during the Middle and Late Permian". Geology. 39 (11): 1059–1062. Bibcode:2011Geo....39.1059C. doi:10.1130/G32230.1.

111. Payne JL, Clapham ME (2012). "End-Permian Mass Extinction in the Oceans: An Ancient Analog for the Twenty-First Century?". Annual Review of Earth and Planetary Sciences. 40: 89–111. Bibcode:2012AREPS..40...89P. doi:10.1146/annurev-earth-042711-105329.

112. Life in the Sea Found Its Fate in a Paroxysm of Extinction New York Times, April 30, 2012.

113. http://www.savethehighseas.org/theproblem/

114. Hutchings, P.A. (1986). "Biological destruction of coral reefs". Coral Reefs. Springer-Verlag. 12 (1): 1–17. Bibcode:1986CorRe...4..239H. doi:10.1007/BF00298083.

115. Losing Our Coral Reefs – Eco Matters – State of the Planet. Blogs.ei.columbia.edu. Retrieved on November 1, 2011.

116. Ritter, Karl (December 8, 2010). −goal-coral-reefs.html "Climate goal may spell end for some coral reefs". Associated Press. Retrieved December 2010. Check date values in: |access-date= (help)

117. Markey, Sean (May 16, 2006). "Global Warming Has Devastating Effect on Coral Reefs, Study Shows". National Geographic News.

118. Abrego D, Ulstrup KE, Willis BL, van Oppen MJH (2008) Species–specific interactions between algal endosymbionts and coral hosts define their bleaching response to heat and light stress. Proc Biol Sci 275: 2273–2282

119. Kleypas, J.A.; Feely, R.A.; Fabry, V.J.; Langdon, C.; Sabine, C.L.; Robbins, L.L. (2006). "Impacts of Ocean Acidification on Coral Reefs and Other Marine Calcifiers: A guide for Future Research" (PDF). National Science Foundation, NOAA, & United States Geological Survey. Retrieved April 7, 2011.

120. Tun, K.; Chou, L.M.; Cabanban, A.; Tuan, V.S.; Philreefs; Yeemin, T.; Suharsono; Sour, K.; Lane, D. (2004). "Status of Coral Reefs, Coral Reef Monitoring and Management in Southeast Asia, 2004". In Wilkinson, C. Status of Coral Reefs of the world: 2004 (PDF). Townsville, Queensland, Australia: Australian Institute of Marine Science. pp. 235–276.

121. "Reefs at Risk Revisited: Southeast Asia" (PDF). World Resources Institute. February 2011. Retrieved January 27, 2014.

122. "Reefs at Risk Revisited". World Resources Institute. February 2011. pp. 45–46.

Retrieved January 27, 2014.

123. Lamb, Joleah; Bette Willis (August 16, 2011). "Using coral disease prevalence to assess the effects of concentrating tourism activities on offshore reefs in a tropical marine park". Conservation Biology. 25 (5): 1044–1052. doi:10.1111/j.1523-1739.2011.01724.x. PMID 21848962.

124. McClanahan, Timothy; Marnane, Michael; Cinner, Joshua E.; Kiene, William E. (2006). "A Comparison of Marine Protected Areas and Alternative Approaches to Coral-Reef Management". Current Biology. 16 (14): 1408–13. doi:10.1016/j.cub.2006.05.062. PMID 16860739.

125. Christie, P. (2004). "Marine protected areas as biological successes and social failures in Southeast Asia". American Fisheries Society Symposium. 2004 (42): 155–164.

126. McClanahan, Timothy; Davies, Jamie; Maina, Joseph (2005). "Factors influencing resource users and managers' perceptions towards marine protected area management in Kenya". Environmental Conservation. 32: 42–49. doi:10.1017/S0376892904001791.

127. Stone, Gregory (January 2011). "Phoenix Rising". National Geographic Magazine.

128. Ewa Magiera; Sylvie Rockel (2 July 2014). "From despair to repair: Dramatic decline of Caribbean corals can be reversed". Retrieved 8 June 2015.

129. Kelly, R.P; Foley; Fisher, WS; Feely, RA; Halpern, BS; Waldbusser, GG; Caldwell, MR; et al. (2011). "Mitigating local causes of ocean acidification with existing laws" (PDF). Science. 332 (6033): 1036–1037. Bibcode:2011Sci...332.1036K. doi:10.1126/science.1203815. PMID 21617060.

130. "Satellites to assess coral reef health". Retrieved 13 December 2014.

131. "A biodiversity strategy for the Great Barrier Reef". Great Barrier Reef Marine Park Authority, Australian Government. Retrieved 20 September 2013.

132. "Great Barrier Reef Climate Change Action Plan 2007–2011" (PDF). Great Barrier Reef Marine Park Authority. 2007.

133. Cinner, Joshua E.; Marnane, Michael J.; McClanahan, Tim R. (2005). "Conservation and community benefits from traditional coral reef management at Ahus Island, Papua New Guinea". Conservation Biology. 19 (6): 1714–1723. doi:10.1111/j.1523-1739.2005.00209.x-i1.

134. "Coral Reef Management, Papua New Guinea". Nasa's Earth Observatory. Retrieved November 2, 2006.

135. Horoszowski-Fridman, YB, Izhaki, I & Rinkevich, B; Izhaki; Rinkevich (2011). "Engineering of coral reef larval supply through transplantation of nursery-farmed gravid colonies". Journal of Experimental Marine Biology and Ecology. 399 (2): 162–166. doi:10.1016/j.jembe.2011.01.005.

136. Pomeroy, RS, Parks, JE and Balboa, CM (2006). "Farming the reef: is aquaculture a solution for reducing fishing pressure on coral reefs?". Marine Policy. 30 (2): 111–130. doi:10.1016/j.marpol.2004.09.001.

137. Rinkevich, B (2008). "Management of coral reefs: We have gone wrong when

neglecting active reef restoration" (PDF). Marine pollution bulletin. 56 (11): 1821–1824. doi:10.1016/j.marpolbul.2008.08.014. PMID 18829052.[permanent dead link]

138. Ferse, SCA (2010). "Poor Performance of Corals Transplanted onto Substrates of Short Durability". Restoration Ecology. 18 (4): 399–407. doi:10.1111/j.1526-100X.2010.00682.x.

139. Superglue used for placement of coral. coralgarden.co.uk (May 8, 2009). Retrieved on November 8, 2011.

140. Needle and thread use with soft coral. coralgarden.co.uk (May 8, 2009). Retrieved on November 8, 2011.

141. Jump up to: a b Sabater, Marlowe G.; Yap, Helen T. (2004). "Long-term effects of induced mineral accretion on growth, survival, and corallite properties of Porites cylindrica Dana" (PDF). Journal of Experimental Marine Biology and Ecology. 311 (2): 355–374. doi:10.1016/j.jembe.2004.05.013.

142. Jokeil, P.L.; Ku'lei, S.R (2004). "Coral Relocation Project in Kaneohe Bay, Oahu, Hawaii: Report on Phase 1" (PDF). University of Hawaii.

143. "Gene Therapy Could Help Corals Survive Climate Change". Scientific American. February 29, 2012.

144. Cobb, K.; Charles, Christopher D.; Cheng, Hai; Edwards, R. Lawrence (2003). "El Nino/Southern Oscillation and tropical Pacific climate during the past millennium" (PDF). Nature. 424 (6946): 271–6. Bibcode:2003Natur.424..271C. doi:10.1038/nature01779. PMID 12867972.

CHAPTER 9

Marine Pollution

Chapter Contents

Marine pollution refers to ocean contamination, especially with man-made waste or byproducts. This is considered to be a severe problem by many environmentalists, since it can have detrimental and lasting effects on the global marine ecosystem. There are several types of marine pollution, each of which has the potential to harm the delicate balance of life.

INTRODUCTION

Marine pollution occurs when harmful, or potentially harmful, effects result from the entry into the ocean of chemicals, particles, industrial, agricultural and residential waste, noise, or the spread of invasive organisms. Eighty percent of marine pollution comes from land. Air pollution is also a contributing factor by carrying off pesticides or dirt into the ocean. Land and air pollution have proven to be harmful to marine life and its habitats.

While marine pollution can be obvious, as with the marine debris shown above, it is often the pollutants that cannot be seen that cause most harm

The pollution often comes from non point sources such as agricultural runoff, wind-blown debris and dust. Nutrient pollution, a form of water pollution, refers to contamination by excessive inputs of nutrients. It is a primary cause of eutrophication of surface waters, in which excess nutrients, usually nitrogen or phosphorus, stimulate algae growth.

Many potentially toxic chemicals adhere to tiny particles which are then taken up by plankton and benthos animals, most of which are either deposit or filter feeders. In this way, the toxins are concentrated upward within ocean food chains. Many particles combine chemically in a manner highly depletive of oxygen, causing estuaries to become anoxic.

When pesticides are incorporated into the marine ecosystem, they quickly become absorbed into marine food webs. Once in the food webs, these pesticides can cause mutations, as well as diseases, which can be harmful to humans as well as the entire food web.

Toxic metals can also be introduced into marine food webs. These can cause a change to tissue matter, biochemistry, behavior, reproduction, and suppress growth in marine life. Moreover, many animal feeds have a high fish meal or fish hydrolysate content. In this way, marine toxins can be transferred to land animals, and appear later in meat and dairy products.

History

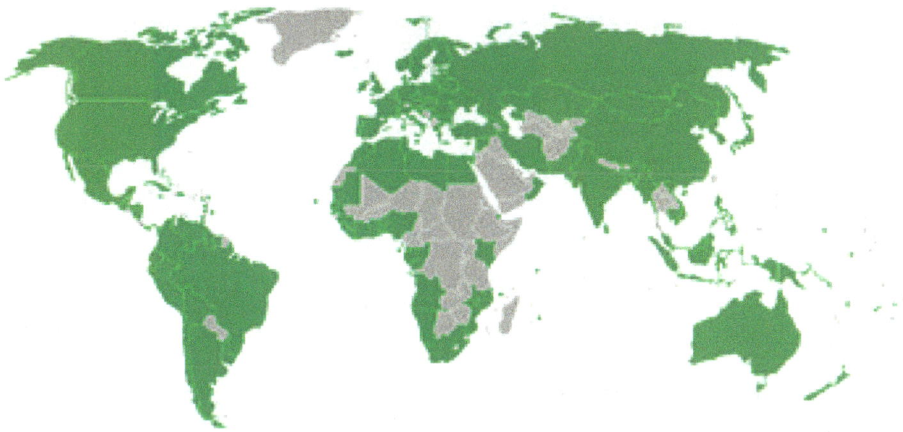

Parties to the MARPOL 73/78 convention on marine pollution

Although marine pollution has a long history, significant international laws to counter it were only enacted in the twentieth century. Marine pollution was a concern during several United Nations Conferences on the Law of the Sea beginning in the 1950s. Most scientists believed that the oceans were so vast that they had unlimited ability to dilute, and thus render pollution harmless.

In the late 1950s and early 1960s, there were several controversies about dumping radioactive waste off the coasts of the United States by companies licensed by the Atomic Energy Commission, into the Irish Sea from the British reprocessing facility at Windscale, and into the Mediterranean Sea by the French Commissariat à l›Energie Atomique. After the Mediterranean Sea controversy, for example, Jacques Cousteau became a worldwide figure in the campaign to stop marine pollution. Marine pollution made further international headlines after the 1967 crash of the oil tanker Torrey Canyon, and after the 1969 Santa Barbara oil spill off the coast of California.

Marine pollution was a major area of discussion during the 1972 United Nations Conference on the Human Environment, held in Stockholm. That year also saw the signing of the Convention on the Prevention of Marine Pollution by Dumping of Wastes and Other Matter, sometimes called the London Convention. The London Convention did not ban marine pollution, but it established black and gray lists for substances to be banned (black) or regulated by national authorities (gray). Cyanide and high-level radioactive waste, for example, were put on the black list. The London Convention applied only to waste dumped from ships, and thus did nothing to regulate waste discharged as liquids from pipelines.

PATHWAYS OF POLLUTION

Septic river

There are many different ways to categorize, and examine the inputs of pollution into our marine ecosystems. Patin (n.d.) notes that generally there are three main types of inputs of pollution into the ocean: direct discharge of waste into the oceans, runoff into the waters due to rain, and pollutants that are released from the atmosphere.

One common path of entry by contaminants to the sea are rivers. The evaporation of water from oceans exceeds precipitation. The balance is restored by rain over the continents entering rivers and then being returned to the sea. The Hudson in New York State and the Raritan in New Jersey, which empty at the northern and southern ends of Staten Island, are a source of mercury contamination of zooplankton (copepods) in the open ocean. The highest concentration in the filter-feeding copepods is not at the mouths of these rivers but 70 miles south, nearer Atlantic City, because water flows close to the coast. It takes a few days before toxins are taken up by the plankton. Pollution is often classed as point source or nonpoint source pollution. Point source pollution occurs when there is a single, identifiable, and localized source of the pollution. An example is directly discharging sewage and industrial waste into the ocean. Pollution such as this occurs particularly in developing nations. Nonpoint source pollution occurs when the pollution comes from ill-defined and diffuse sources. These can be difficult to regulate. Agricultural runoff and wind blown debris are prime examples.

Direct Discharge

Acid mine drainage in the Rio Tinto River.

See also: Sewerage, Industrial waste, and Environmental issues with mining

Pollutants enter rivers and the sea directly from urban sewerage and industrial waste discharges, sometimes in the form of hazardous and toxic wastes.

Inland mining e.g. copper and gold is another source of marine pollution. Most of the pollution is simply soil, which ends up in rivers flowing to the sea. However, some minerals discharged in the course of the mining can cause problems, such as copper, a common industrial pollutant, which can interfere with the life history and development of coral polyps. Mining has a poor environmental track record. For example, according to the United States Environmental Protection Agency, mining has contaminated portions of the headwaters of over 40% of watersheds in the western continental US. Much of this pollution finishes up in the sea.

Land Runoff

Surface runoff from farming, as well as urban runoff and runoff from the construction of roads, buildings, ports, channels, and harbours, can carry soil and particles laden with carbon, nitrogen, phosphorus, and minerals. This nutrient-rich water can cause fleshy algae and phytoplankton to thrive in coastal areas; known as algal blooms, which have the potential to create hypoxic conditions by using all available oxygen.

Polluted runoff from roads and highways can be a significant source of water pollution in coastal areas. About 75% of the toxic chemicals that flow into Puget Sound are carried by stormwater that runs off paved roads and driveways, rooftops, yards and other developed land.

Ship Pollution

A cargo ship pumps ballast water over the side

Ships can pollute waterways and oceans in many ways. Oil spills can have devastating effects. While being toxic to marine life, polycyclic aromatic hydrocarbons (PAHs), found in crude oil, are very difficult to clean up, and last for years in the sediment and marine environment.

Oil spills are probably the most emotive of marine pollution events. However, while a tanker wreck may result in extensive newspaper headlines, much of the oil in the world's seas comes from other smaller sources, such as tankers discharging ballast water from oil tanks used on return ships, leaking pipelines or engine oil disposed of down sewers.

Discharge of cargo residues from bulk carriers can pollute ports, waterways and oceans. In many instances vessels intentionally discharge illegal wastes despite foreign and domestic regulation prohibiting such actions. It has been estimated that container ships lose over 10,000 containers at sea each year (usually during storms). Ships also create noise pollution that disturbs natural wildlife, and water from ballast tanks can spread harmful algae and other invasive species.

Ballast water taken up at sea and released in port is a major source of unwanted exotic marine life. The invasive freshwater zebra mussels, native to the Black, Caspian, and Azov seas, were probably transported to the Great Lakes via ballast water from a transoceanic vessel. Meinesz believes that one of the worst cases of a single invasive species causing harm to an ecosystem can be attributed to a seemingly harmless jellyfish. *Mnemiopsis leidyi*, a species of comb jellyfish that spread so it now inhabits estuaries in many parts of the world. It was first introduced in 1982, and thought to have been transported to the Black Sea in a ship's ballast water. The population of the jellyfish shot up exponentially and, by 1988, it was wreaking havoc upon the local fishing industry. "The anchovy catch fell from 204,000 tons in 1984 to 200 tons in 1993; sprat from 24,600 tons in 1984 to 12,000 tons in 1993; horse mackerel from 4,000 tons in 1984 to zero in 1993." Now that the jellyfish have exhausted the zooplankton, including fish larvae, their numbers have fallen dramatically, yet they continue to maintain a stranglehold on the ecosystem.

Invasive species can take over once occupied areas, facilitate the spread of new diseases, introduce new genetic material, alter underwater seascapes and jeopardize the ability of native species to obtain food. Invasive species are responsible for about $138 billion annually in lost revenue and management costs in the US alone.

Atmospheric Pollution

Another pathway of pollution occurs through the atmosphere. Wind blown dust and debris, including plastic bags, are blown seaward from landfills and other areas. Dust from the Sahara moving around the southern periphery of the subtropical ridge moves into the Caribbean and Florida during the warm season as the ridge builds and moves northward through the subtropical Atlantic. Dust can also be attributed to a global transport from the Gobi and Taklamakan deserts across Korea, Japan, and the Northern Pacific to the Hawaiian Islands. Since 1970, dust outbreaks have worsened due to periods of drought in Africa. There is a large variability in dust transport to the Caribbean and Florida from year to year; however, the flux is greater during positive phases of the North Atlantic Oscillation. The USGS links dust events to a decline in the health of coral reefs across the Caribbean and Florida, primarily since the 1970s.

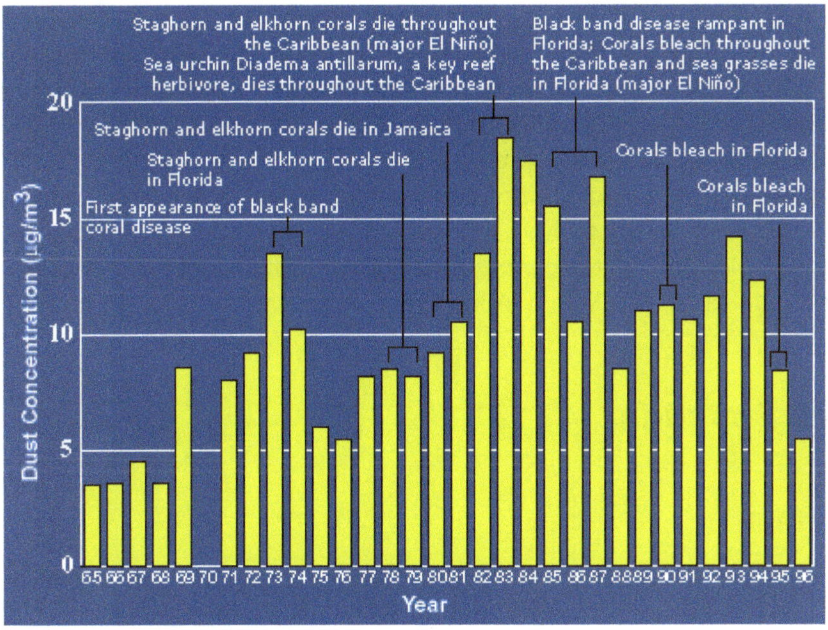

Graph linking atmospheric dust to various coral deaths across the Caribbean Sea and Florida

Climate change is raising ocean temperatures and raising levels of carbon dioxide in the atmosphere. These rising levels of carbon dioxide are acidifying the oceans. This, in turn, is altering aquatic ecosystems and modifying fish distributions,with impacts on the sustainability of fisheries and the livelihoods of the communities that depend on them. Healthy ocean ecosystems are also important for the mitigation of climate change.

Deep sea Mining

Deep sea mining is a relatively new mineral retrieval process that takes place on the ocean floor. Ocean mining sites are usually around large areas of polymetallic nodules or active and extinct hydrothermal vents at about 1,400–3,700 meters below the ocean's surface. The vents create sulfide deposits, which contain precious metals such as silver, gold, copper, manganese, cobalt, and zinc. The deposits are mined using either hydraulic pumps or bucket systems that take ore to the surface to be processed. As with all mining operations, deep sea mining raises questions about environmental damages to the surrounding areas

Because deep sea mining is a relatively new field, the complete consequences of full scale mining operations are unknown. However, experts are certain that removal of parts of the sea floor will result in disturbances to the benthic layer, increased toxicity of the water column and sediment plumes from tailings. Removing parts of the sea floor disturbs the habitat of benthic organisms, possibly, depending on the type of mining and location, causing permanent disturbances. Aside from direct impact of mining the area, leakage, spills and corrosion would alter the mining area's chemical makeup.

Among the impacts of deep sea mining, sediment plumes could have the greatest impact. Plumes are caused when the tailings from mining (usually fine particles) are dumped back into the ocean, creating a cloud of particles floating in the water. Two types of plumes occur: near bottom plumes and surface plumes. Near bottom plumes occur when the tailings are pumped back down to the mining site. The floating particles increase the turbidity, or cloudiness, of the water, clogging filter-feeding apparatuses used by benthic organisms. Surface plumes cause a more serious problem. Depending on the size of the particles and water currents the plumes could spread over vast areas. The plumes could impact zooplankton and light penetration, in turn affecting the food web of the area.

Types of Pollution

Acidification

Island with fringing reef in the Maldives. Coral reefs are dying around the world

The oceans are normally a natural carbon sink, absorbing carbon dioxide from the atmosphere. Because the levels of atmospheric carbon dioxide are increasing, the oceans are becoming more acidic. The potential consequences of ocean acidification are not fully understood, but there are concerns that structures made of calcium carbonate may become vulnerable to dissolution, affecting corals and the ability of shellfish to form shells.

Oceans and coastal ecosystems play an important role in the global carbon cycle and have removed about 25% of the carbon dioxide emitted by human activities between 2000 and 2007 and about half the anthropogenic CO_2 released since the start of the industrial revolution. Rising ocean temperatures and ocean acidification means that the capacity of the ocean carbon sink will gradually get weaker, giving rise to global concerns expressed in the Monaco and Manado Declarations.

A report from NOAA scientists published in the journal Science in May 2008 found that large amounts of relatively acidified water are upwelling to within four miles of the

Pacific continental shelf area of North America. This area is a critical zone where most local marine life lives or is born. While the paper dealt only with areas from Vancouver to northern California, other continental shelf areas may be experiencing similar effects.

A related issue is the methane clathrate reservoirs found under sediments on the ocean floors. These trap large amounts of the greenhouse gas methane, which ocean warming has the potential to release. In 2004 the global inventory of ocean methane clathrates was estimated to occupy between one and five million cubic kilometres. If all these clathrates were to be spread uniformly across the ocean floor, this would translate to a thickness between three and fourteen metres. This estimate corresponds to 500–2500 gigatonnes carbon (Gt C), and can be compared with the 5000 Gt C estimated for all other fossil fuel reserves.

Eutrophication

Polluted lagoon

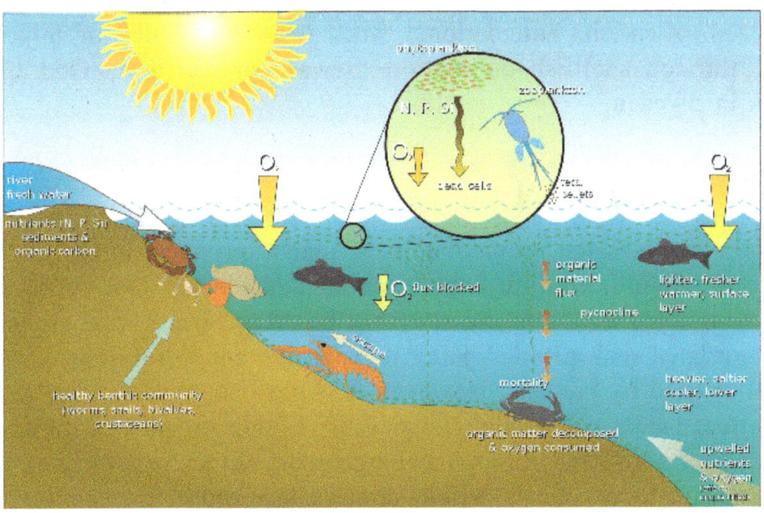

Effect of eutrophication on marine benthic life

Eutrophication is an increase in chemical nutrients, typically compounds containing nitrogen or phosphorus, in an ecosystem. It can result in an increase in the ecosystem›s primary productivity (excessive plant growth and decay), and further effects including lack of oxygen and severe reductions in water quality, fish, and other animal populations.

The biggest culprit are rivers that empty into the ocean, and with it the many chemicals used as fertilizers in agriculture as well as waste from livestock and humans. An excess of oxygen depleting chemicals in the water can lead to hypoxia and the creation of a dead zone.

Estuaries tend to be naturally eutrophic because land-derived nutrients are concentrated where runoff enters the marine environment in a confined channel. The World Resources Institute has identified 375 hypoxic coastal zones around the world, concentrated in coastal areas in Western Europe, the Eastern and Southern coasts of the US, and East Asia, particularly in Japan. In the ocean, there are frequent red tide algae blooms that kill fish and marine mammals and cause respiratory problems in humans and some domestic animals when the blooms reach close to shore.

In addition to land runoff, atmospheric anthropogenic fixed nitrogen can enter the open ocean. A study in 2008 found that this could account for around one third of the ocean's external (non-recycled) nitrogen supply and up to three per cent of the annual new marine biological production. It has been suggested that accumulating reactive nitrogen in the environment may have consequences as serious as putting carbon dioxide in the atmosphere.

One proposed solution to eutrophication in estuaries is to restore shellfish populations, such as oysters. Oyster reefs remove nitrogen from the water column and filter out suspended solids, subsequently reducing the likelihood or extent of harmful algal blooms or anoxic conditions. Filter feeding activity is considered beneficial to water quality by controlling phytoplankton density and sequestering nutrients, which can be removed from the system through shellfish harvest, buried in the sediments, or lost through denitrification. Foundational work toward the idea of improving marine water quality through shellfish cultivation to was conducted by Odd Lindahl et al., using mussels in Sweden.

PLASTIC POLLUTION

Plastic pollution involves the accumulation of plastic products in the environment that adversely affects wildlife, wildlife habitat, or humans.Plastics that act as pollutants are categorized into micro-, meso-, or macrodebris, based on size. The prominence of plastic pollution is correlated with plastics being inexpensive and durable, which lends to high levels of plastics used by humans. However, it is slow to degrade. Plastic pollution can unfavorably affect lands, waterways and oceans. Living organisms, particularly marine animals, can also be affected through entanglement, direct ingestion of plastic waste, or through exposure to chemicals within plastics that cause interruptions in biological functions. Humans are also affected by plastic pollution, such as through the disruption of the thyroid hormone axis or hormone levels. In the UK alone, more than 5 million tonnes of plastic are consumed each year, of which an estimated mere 24% makes it into recycling systems. That leaves a remaining 3.8 million tons of waste, destined for landfills. Plastic reduction efforts have occurred in some areas in attempts to reduce plastic consumption and pollution and promote plastic recycling.

Plastic waste at Coco Beach in India.

Types of Plastic Debris

There are three major forms of plastic that contribute to plastic pollution: microplastics as well as mega- and macro-plastics. Mega- and micro plastics have accumulated in highest densities in the Northern Hemisphere, concentrated around urban centers and water fronts. Plastic can be found off the coast of some islands because of currents carrying the debris. Both mega- and macro-plastics are found in packaging, footwear, and other domestic items that have been washed off of ships or discarded in landfills. Fishing-related items are more likely to be found around remote islands. These may also be referred to as micro-, meso-, and macro debris.

Plastic debris is categorized as either primary or secondary. Primary plastics are in their original form when collected. Examples of these include bottle caps, cigarette butts, and microbeads. Secondary plastics, on the other hand, account for smaller plastics that have resulted from the degradation of primary plastics.

Microdebris

Microdebris are plastic pieces between 2 μm and 5 mm in size. Plastic debris that starts off as meso- or macrodebris can become microdebris through degradation and collisions that break it down into smaller pieces. Microdebris is more commonly referred to as nurdles. Nurdles are recycled to make new plastic items, but they easily end up released into the environment during production because of their small size. They often end up in ocean waters through rivers and streams.Microdebris that come from cleaning and cosmetic products are also referred to as scrubbers. Because microdebris and scrubbers are so small in size, filter-feeding organisms often consume them. A 2004 study by Richard Thompson from the University of Plymouth, UK, found a great amount of microdebris on the beaches and waters in Europe, the Americas, Australia, Africa, and Antarctica. Thompson and his associates found that plastic pellets from both domestic and industrial sources were being broken down into much smaller plastic pieces, some having a diameter smaller than human hair. If not ingested, this microdebris floats instead of being absorbed into the marine environment. Thompson predicts there may be 300,000 plastic items/km^2 of sea surface and 100,000 plastic particles/km^2 of seabed.

Macrodebris

Plastic debris is categorized as macrodebris when it is larger than 20 mm. These include items such as plastic grocery bags.Macrodebris are often found in ocean waters, and can have a serious impact on the native organisms. Fishing nets have been prime pollutants. Even after they have been abandoned, they continue to trap marine organisms and other plastic debris. Eventually, these abandoned nets become too difficult to remove from the water because they become too heavy, having grown in weight up to 6 tons.

Decomposition of Plastics

Plastics themselves contribute to approximately 10% of discarded waste. Many kinds of plastics exist depending on their precursors and the method for their polymerization. Depending on their chemical composition, plastics and resins have varying properties related to contaminant absorption and adsorption. Polymer degradation takes much longer as a result of haline environments and the cooling effect of the sea. These factors contribute to the persistence of plastic debris in certain environments. Recent studies have shown that plastics in the ocean decompose faster than was once thought, due to exposure to sun, rain, and other environmental conditions, resulting in the release of toxic chemicals such as bisphenol A. However, due to the increased volume of plastics in the ocean, decomposition is slowed down. The Marine Conservancy has predicted the decomposition rates of several plastic products. It is estimated that a foam plastic cup will take 50 years, a plastic beverage holder will take 400 years, disposable diaper will take 450 years, and fishing line will take 600 years to degrade.

Persistent Organic Pollutants

It is estimated that global production of plastics is approximately 225 mt/yr. Their abundance has been found to transport persistent organic pollutants, also known as POPs. These pollutants have been linked to an increased distribution of algae associated with red tides.

Effects on the Environment

The distribution of plastic debris is highly variable as a result of certain factors such as wind and ocean currents, coastline geography, urban areas, and trade routes. Human population in certain areas also plays a large role in this. Plastics are more likely to be found in enclosed regions such as the Caribbean. Plastic pollution, more so in the forms of macro- and mega-plastics, potentially serves as a means of distribution of organisms to remote coasts that are not their native environments. This could potentially increase the variability and dispersal of organisms in specific areas that are less biologically diverse. Plastics can also be used as vectors for chemical contaminants such as persistent organic pollutants and heavy metals.

Land

Chlorinated plastic can release harmful chemicals into the surrounding soil, which can then seep into groundwater or other surrounding water sources and also the ecosystem. This can cause serious harm to the species that drink the water.

Landfill areas contain many different types of plastics. In these landfills, there are many microorganisms which speed up the biodegradation of plastics. The microorganisms include bacteria such as Pseudomonas, nylon-eating bacteria, and Flavobacteria. These bacteria break down nylon through the activity of the nylonase enzyme. When biodegradable plastics are broken down, methane is released, which is a very powerful greenhouse gas that contributes significantly to global warming.

Ocean

In 2012, it was estimated that there was approximately 165 million tons of plastic pollution in the world's oceans. One type of plastic that is of concern in terms of ocean plastic pollution is nurdles. Nurdles are manufactured plastic pellets (a type of microplastic) used in the creation of plastic products and are often shipped via cargo ship. A significant amount of nurdles is spilled into oceans, and it has been estimated that globally, around 10% of beach litter consists of nurdles. Plastics in oceans typically degrade within a year, but not entirely. In the process, toxic chemicals such as bisphenol A and polystyrene can leach into waters from some plastics. Polystyrene pieces and nurdles are the most common types of plastic pollution in oceans, and combined with plastic bags and food containers make up the majority of oceanic debris. One study estimated that there are more than 5 trillion plastic pieces (defined into the four classes of small microplastics, large microplastics, meso- and macroplastics) afloat at sea.

The litter that is being delivered into the oceans is toxic to marine life, and humans. The toxins that are components of plastic include Diethylhexyl phthalate, which is a toxic carcinogen, as well as lead, cadmium and mercury (Andrews 2012).

Plankton, fish and ultimately the human race through the food chain, ingest these highly toxic carcinogens and chemicals. Consuming the fish that contain these toxins can cause an increase in cancer, immune disorders and birth defects.

The majority of the litter near and in the ocean is made up of plastics. According to Dr. Marcus Eriksen of 5 Gyres, there are 5.25 trillion particles of plastic pollution that weigh as much as 270,000 tons (2016). This plastic is taken by the ocean currents and accumulates in large vortexes known as gyres. The majority of the gyres become pollution dumps filled with plastic.

Ocean-based Sources of Ocean Plastic Pollution

Almost 90% of plastic debris that pollutes ocean water, which translates to 5.6 million tons, comes from ocean-based sources. Merchant ships expel cargo, sewage, used medical equipment, and other types of waste that contain plastic into the ocean. Naval and research vessels also eject waste and military equipment that are deemed unnecessary. Pleasure crafts also release fishing gear and other types of waste. These different ships do not have enough storage space to keep these pollutants on the ship, and thus they are discarded. These plastic items can also accidentally end up in the water through negligent handling. The largest ocean-based source of plastic pollution is discarded fishing gear, responsible for up to 90% of plastic debris in some areas. This equipment includes a variety of traps and nets.

Land-based sources of Ocean Plastic Pollution

A little over 10% of plastic debris in ocean water comes from land-based sources, responsible for 0.8 million tons every year.A source that has caused concern is landfills. Most waste in the form of plastic in landfills are single-use items such as packaging. Discarding plastics this way leads to accumulation. Although disposing of plastic waste in landfills has less of a gas emission risk than disposal through incineration, the former has space limitations. Another concern is that the liners acting as protective layers between the landfill and environment can break, thus leaking toxins and contaminating the nearby soil and water. Landfills located near oceans often contribute to ocean debris because content is easily swept up and transported to the sea by wind or small waterways like rivers and streams. Marine debris can also result from sewage water that has not been efficiently treated, which is eventually transported to the ocean through rivers. Plastic items that have been improperly discarded can also be carried to oceans through storm waters.

Plastic pollution in the Pacific Gyre

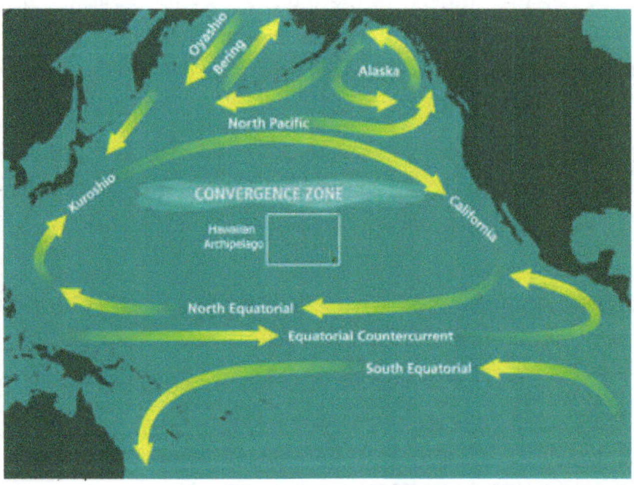

North Pacific Subtropical Convergence Zone

In the Pacific Gyre, specifically 20°N-40°N latitude, large bodies with floating marine debris can be found. Models of wind patterns and ocean currents indicate that the plastic waste in the northern Pacific is particularly dense where the Subtropical Convergence Zone (STCZ), 23°N-37°N latitude, meets a southwest-northeast line, found north of the Hawaiian archipelago.

In the Pacific, there are two mass buildups: the western garbage patch and the eastern garbage patch, the former off the coast of Japan and the latter between Hawaii **and** California. The two garbage patches are both part of the great Pacific garbage patch, and are connected through a section of plastic debris off the northern coast of the Hawaiian islands. It is approximated that these garbage patches contain 100 million tons of debris. The waste is not compact, and although most of it is near the surface of the pacific, it can be found up to more than 100 feet deep in the water.

Effects on Animals

Plastic pollution has the potential to poison animals, which can then adversely affect human food supplies. Plastic pollution has been described as being highly detrimental to large marine mammals, described in the book Introduction to Marine Biology as posing the "single greatest threat" to them. Some marine species, such as sea turtles, have been found to contain large proportions of plastics in their stomach. When this occurs, the animal typically starves, because the plastic blocks the animal`s digestive tract, which block the passage of air and kill them Marine mammals sometimes become entangled in plastic products such as nets, which can harm or kill them.

Entanglement

Sea turtle entangled in a ghost net

Entanglement in plastic debris has been responsible for the deaths of many marine organisms, such as fish, seals, turtles, and birds. These animals get caught in the debris and end up suffocating or drowning. Because they are unable to untangle themselves, they also die from starvation or from their inability to escape predators.Being entangled also often results in severe lacerations and ulcers. In a 2006 report known as *Plastic Debris in the World's Oceans*, it was estimated that at least 267 different animal species have suffered from entanglement and ingestion of plastic debris. It has been estimated that over 400,000 marine mammals perish annually due to plastic pollution in oceans. Marine organisms get caught in discarded fishing equipment, such as ghost nets. Ropes and nets used to fish are often made of synthetic materials such as nylon, making fishing equipment more durable and buoyant. These organisms can also get caught in circular plastic packaging materials, and if the animal continues to grow in size, the plastic can cut into their flesh. Equipment such as nets can also drag along the seabed, causing damage to coral reefs.

Ingestion

Marine animals

Sea turtles are affected by plastic pollution. Some species are consumers of jelly fish, but often mistake plastic bags for their natural prey. This plastic debris can kill the sea turtle by obstructing the esophagus. So too are whales; large amounts of plastics have been found in the stomachs of beached whales

Some of the tiniest bits of plastic are being consumed by small fish, in a part of the pelagic zone in the ocean called the Mesopelagic zone, which is 200 to 1000 m below the ocean surface, and completely dark. Not much is known about these fish, other than that there are many of them. They hide in the darkness of the ocean, avoiding predators and then swimming to the ocean's surface at night to feed. Plastics found in the stomachs of these fish were collected during Malaspina's circumnavigation, a research project that studies the impact of global change on the oceans. The most popular mesopelagic fish is the lantern fish. It resides in the central ocean gyres, a large system of rotating ocean currents. Since lantern fish serve as a primary food source for the fish that consumers purchase, including tuna and swordfish, the plastics they ingest become part of the food chain. The lantern fish is one of the main bait fish in the ocean, and it eats large amounts of plastic fragments, which in turn will not make them nutritious enough for other fish to consume.

Birds

Plastic pollution does not only affect animals that live solely in oceans. Seabirds are also greatly affected. In 2004, it was estimated that seagulls in the North Sea had an average of thirty pieces of plastic in their stomachs. Seabird's often mistake trash floating on the ocean's surface as prey. Their food sources often has already ingested plastic debris, thus transferring the plastic from prey to predator. Ingested trash can obstruct and physically damage a bird's digestive system, reducing its digestive ability and can lead to malnutrition, starvation, and death. Toxic chemicals called polychlorinated biphenyls (PCBs) also become concentrated on the surface of plastics at sea and are released after seabirds eat them. These chemicals can accumulate in body tissues and have serious lethal effects on a bird's reproductive ability, immune system, and hormone balance. Floating plastic debris can produce ulcers, infections and lead to death. Marine plastic pollution can even reach birds that have never been at the sea. Parents may accidentally feed their nestlings plastic, mistaking it for food. Seabird chicks are the most vulnerable to plastic ingestion since they can't regurgitate like the adult seabirds.

After the initial observation that many of the beaches in New Zealand had high concentrations of plastic pellets, further studies found that different species of prion ingest the plastic debris. Hungry prions mistook these pellets for food, and these particles were found intact within the birds' gizzards and proventriculi. Pecking marks similar to those made by northern fulmars in cuttlebones have been found in plastic debris, such as styrofoam, on the beaches on the Dutch coast, showing that this species of bird also mistakes plastic debris for food.

An estimate of 1.5 million Laysan albatrosses, which inhabit Midway Atoll, all have plastics in their digestive system. Midway Atoll is halfway between Asia and North America, and north of the Hawaiian archipelago. It's a remote location, and the plastic blockage has proven deadly to these birds. These seabirds choose red, pink, brown and blue plastic pieces because of the similarities they share with their natural food source. On the shore, thousands of birds corpses can be seen with plastic remaining where the stomach once was. The durability of the plastics is visible amongst the remains. In some instances, the plastic piles are still present while the bird`s corpse has decayed. As a result of plastic ingestion, the digestive tract can be blocked resulting in starvation. The windpipe can also be blocked, which results in suffocation. The debris can also accumulate in the animal's gut, and give them a false sense of fullness which would also result in starvation.

Effects on animals

Similar to humans, animals exposed to plasticizers can experience developmental defects. Specifically, sheep have been found to have lower birth weights when prenatally exposed to bisphenol A (BPA). Exposure to BPA can shorten the distance between the eyes of a tadpole. It can also stall development in frogs and can result in a decrease in body length. In different species of fish, exposure can stall egg hatching and result in a decrease in body weight, tail length, and body length.

Effects on humans

Due to the use of chemical additives during plastic production, plastics have potentially harmful effects that could prove to be carcinogenic or promote endocrine disruption. Some of the additives are used as phthalate plasticizers and brominated flame retardants. Through biomonitoring, chemicals in plastics, such as BPA and phthalates, have been identified in the human population. Humans can be exposed to these chemicals through the nose, mouth, or skin. Although the level of exposure varies depending on age and geography, most humans experience simultaneous exposure to many of these chemicals. Average levels of daily exposure are below the levels deemed to be safe, but more research needs to be done on the effects of low dose exposure on humans. A lot is unknown on how severely humans are physically affected by these chemicals. Some of the chemicals used in plastic production can cause dermatitis upon contact with human skin. In many plastics, these toxic chemicals are only used in trace amounts, but significant testing is often required to ensure that the toxic elements are contained within the plastic by inert material or polymer. It can also affect humans in which it may create an eyesore that interferes with enjoyment of the natural environment.

Clinical Significance

Due to the pervasiveness of plastic products, most of the human population is constantly exposed to the chemical components of plastics. 95% of adults in the United States have had detectable levels of BPA in their urine. Exposure to chemicals such as BPA have been correlated with disruptions in fertility, reproduction, sexual maturation, and other health effects. Specific phthalates have also resulted in similar biological effects.

Thyroid Hormone Axis

Bisphenol A affects gene expression related to the thyroid hormone axis, which affects biological functions such as metabolism and development. BPA can decrease thyroid hormone receptor (TR) activity by increasing TR transcriptional corepressor activity. This then decreases the level of thyroid hormone binding proteins that bind to triiodothyronine. By affecting the thyroid hormone axis, BPA expoure can lead to hypothyroidism.

Sex Hormones

BPA can disrupt normal, physiological levels of sex hormones. It does this by binding to globulins that normally bind to sex hormones such as androgens and estrogens, leading to the disruption of the balance between the two. BPA can also affect the metabolism or the catabolism of sex hormones. It often acts as an antiandrogen or as an estrogen, which can cause disruptions in gonadal development and sperm production.

Reduction Efforts

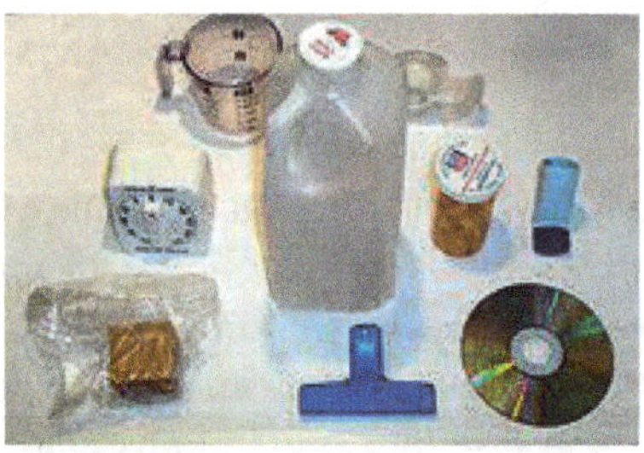

Household items made of various types of plastic.

Efforts to reduce the use of plastics and to promote plastic recycling have occurred. Some supermarkets charge their customers for plastic bags, and in some places more efficient reusable or biodegradable materials are being used in place of plastics. Some communities and businesses have put a ban on some commonly used plastic items, such as bottled water and plastic bags.

Biodegradable and degradable plastics

The use of biodegradable plastics has been shown to have many advantages and disadvantages. Biodegradables are biopolymers that degrade in industrial composters. Biodegradables do not degrade as efficiently in domestic composters, and during this slower process, methane gas may be emitted.

There are also other types of degradable materials that are not considered to be biopolymers, because they are oil-based, similar to other conventional plastics. These plastics are made to be more degradable through the use of different additives, which help them degrade when exposed to UV rays or other physical stressors. However, biodegradation-promoting additives for polymers have been shown not to significantly increase biodegradation.

Although biodegradable and degradable plastics have helped reduce plastic pollution, there are some drawbacks. One issue concerning both types of plastics is that they do not break down very efficiently in natural environments. There, degradable plastics that are oil-based may break down into smaller fractions, at which point they do not degrade further.

Incineration

Up to 60% of used, plastic medical equipment is incinerated rather than deposited in a landfill as a precautionary measure to lessen the transmission of disease. This has allowed for a large decrease in the amount of plastic waste that stems from medical equipment. If plastic waste is not incinerated and disposed of properly, a harmful amount of toxins can be released and dispersed as a gas through air or as ash through air and waterways. Many studies have been done concerning the gaseous emissions that result from the incineration process.

Policy

Agencies such as the Environmental Protection Agency and the Food and Drug Administration often do not assess the safety of new chemicals until after a negative side effect is shown. Once they suspect a chemical may be toxic, it is studied to determine the human reference dose, which is determined to be the lowest observable adverse effect level. During these studies, a high dose is tested to see if it causes any adverse health effects, and if it does not, lower doses are considered to be safe as well. This does not take into account the fact that with some chemicals found in plastics, such as BPA, lower doses can have a discernible effect. Even with this often complicated evaluation process, policies have been put into place in order to help alleviate plastic pollution and its effects. Government regulations have been implemented that ban some chemicals from being used in specific plastic products. In Canada, the United States, and the European Union, BPA has been banned from being incorporated in the production of baby bottles and children's cups, due to health concerns and the higher vulnerability of younger children to the effects of BPA. Taxes have been established in order to discourage specific ways of managing plastic waste. The landfill tax, for example, creates an incentive to choose to recycle plastics rather than contain them in landfills, by making the latter more expensive. There has also been a standardization of the types of plastics that can be considered compostable. The European Norm EN 13432, which was set by the European Committee for Standardization (CEN), lists the standards that plastics must meet, in terms of compostability and biodegradability, in order to officially be labeled as compostable.

Institutional Arrangements in Canada

The Canadian federal government formed a current institution that protects marine areas; this includes the mitigation of plastic pollution. In 1997, Canada adopted legislation for oceans management and passed the Oceans Act. Federal governance, Regional Governance, and Aboriginal Peoples are the actors involved in the process of decision-making and implementation of the decision. The Regional Governance bodies are federal, provincial, and territorial government agencies that hold responsibilities of the marine environment. Aboriginal Peoples in Canada have treaty and non-treaty rights related to ocean activities. According to the Canadian government, they respect these rights and work with Aboriginal groups in oceans management activities.

With the Oceans Act made legal, Canada made a commitment to conserve and protect the oceans. The Ocean Acts' underlying principle is sustainable development, precautionary and integrated management approach to ensure that there is a comprehensive understanding in protecting marine areas. In the integrated management approach, the Oceans Act designates federal responsibility to the Minister of Fisheries and Oceans Canada for any new and emerging ocean-related activities.The Act encourages collaboration and coordination within the government that unifies interested parties. Moreover, the Oceans Act engages any Canadians who are interested in being informed of the decision-making regarding ocean environment.

In 2005, federal organizations developed the Federal Marine Protected Areas Strategy. This strategy is a collaborative approach implemented by Fisheries and Oceans Canada, Parks Canada, and Environment Canada to plan and manage federal marine protected areas. The federal marine protected areas work with Aboriginal groups, industries, academia, environmental groups, and NGOs to strengthen marine protected areas. The federal marine protected areas network consists of three core programs: Marine Protected Areas, Marine Wildlife Areas, and National Marine Conservation Areas. The MPA is a program to be noted because it is significant in protecting ecosystems from the effects of industrial activities. The MPA guiding principles are Integrated Management, Ecosystem-Based Management approach, Adaptive Management Approach, Precautionary Principle, and Flexible Management Approach. All five guiding principles are used collectively and simultaneously to collaborate and respect legislative mandates of individual departments, to use scientific knowledge and traditional ecological knowledge (TEK) to manage human activities, to monitor and report on programs to meet conservation objectives of MPAs, to use best available information in the absence of scientific certainty, and to maintain a balance between conservation needs and sustainable development objectives.

Collection

The two common forms of waste collection include curbside collection and the use of drop-off recycling centers. About 87 percent of the population in the U.S.A. (273 million people) have access to curbside and drop-off recycling centers. In curbside collection, which is available to about 63 percent of the U.S.A. population (193 million people), people place designated plastics in a special bin to be picked up by a public or private hauling company. Most curbside programs collect more than one type of plastic resin; usually both PETE and HDPE. At drop-off recycling centers, which are available to 68 percent of the U.S.A. population (213 million people), people take their recyclables to a centrally located facility. Once collected, the plastics are delivered to a materials recovery facility (MRF) or handler for sorting into single-resin streams to increase product value. The sorted plastics are then baled to reduce shipping costs to reclaimers.

There are varying rates of recycling per type of plastic, and in 2011, the overall plastic recycling rate was approximately 8% in the United States. Approximately 2.7 million tons of plastics were recycled in the U.S. in 2011. Some plastics are recycled more than others; in 2011 29 % of HDPE bottles and 29 % of PET bottles and jars were recycled.

Non-Usage/ Reduction in Usage

The Ministry of Drinking Water and Sanitation, Government of India, has requested various governmental departments to avoid the usage of plastic bottles to provide drinking water during governmental meetings etc., and instead, to make arrangements for providing drinking water that do not result in the generation of plastic waste. The state of Sikkim has restricted the usage of plastic water bottles (in government functions and meetings) and styrofoam products. The state of Bihar has banned the usage of plastic water bottles in governmental meetings. The 2015 National Games of India, organised in Thiruvananthapuram, was associated with green protocols. This was initiated by Suchitwa Mission that aimed for "zero-waste" venues. To make the event "disposable-free", there was ban on the usage of disposable water bottles. The event witnessed the usage of reusable tableware and stainless steel tumblers. Athletes were provided with refillable steel flasks. It is estimated that these green practices stopped the generation of 120 metric tonnes of disposable waste.

Toxins

Apart from plastics, there are particular problems with other toxins that do not disintegrate rapidly in the marine environment. Examples of persistent toxins are PCBs, DDT, TBT, pesticides, furans, dioxins, phenols, and radioactive waste. Heavy metals are metallic chemical elements that have a relatively high density and are toxic or poisonous at low concentrations. Examples are mercury, lead, nickel, arsenic and cadmium. Such toxins can accumulate in the tissues of many species of aquatic life in a process called bioaccumulation. They are also known to accumulate in benthic environments, such as estuaries and bay muds: a geological record of human activities of the last century.

Specific examples

- Chinese and Russian industrial pollution such as phenols and heavy metals in the Amur River have devastated fish stocks and damaged its estuary soil.
- Wabamun Lake in Alberta, Canada, once the best whitefish lake in the area, now has unacceptable levels of heavy metals in its sediment and fish.
- Acute and chronic pollution events have been shown to impact southern California kelp forests, though the intensity of the impact seems to depend on both the nature of the contaminants and duration of exposure.
- Due to their high position in the food chain and the subsequent accumulation of heavy metals from their diet, mercury levels can be high in larger species such as bluefin and albacore. As a result, in March 2004 the United States FDA issued guidelines recommending that pregnant women, nursing mothers and children limit their intake of tuna and other types of predatory fish.
- Some shellfish and crabs can survive polluted environments, accumulating heavy metals or toxins in their tissues. For example, mitten crabs have a remarkable ability to survive in highly modified aquatic habitats, including polluted waters. The farming and harvesting of such species needs careful management if they are to be used as a food.
- Surface runoff of pesticides can alter the gender of fish species genetically, transforming male into female fish.
- Heavy metals enter the environment through oil spills – such as the Prestige oil spill on the Galician coast – or from other natural or anthropogenic sources.
- In 2005, the ‹Ndrangheta, an Italian mafia syndicate, was accused of sinking at least 30 ships loaded with toxic waste, much of it radioactive. This has led to widespread investigations into radioactive-waste disposal rackets.
- Since the end of World War II, various nations, including the Soviet Union, the United Kingdom, the United States, and Germany, have disposed of chemical weapons in the Baltic Sea, raising concerns of environmental contamination.

Underwater Noise.

See also: Noise pollution, Acoustic ecology, The Effects of Noise on Aquatic Life, and Marine mammals and sonar

Marine life can be susceptible to noise or the sound pollution from sources such as passing ships, oil exploration seismic surveys, and naval low-frequency active sonar. Sound travels more rapidly and over larger distances in the sea than in the atmosphere. Marine animals, such as cetaceans, often have weak eyesight, and live in a world largely defined by acoustic information. This applies also to many deeper sea fish, who live in a world of darkness. Between 1950 and 1975, ambient noise at one location in the Pacific Ocean increased by about ten decibels (that is a tenfold increase in intensity).

Noise also makes species communicate louder, which is called the Lombard vocal response. Whale songs are longer when submarine-detectors are on. If creatures don't "speak" loud enough, their voice can be masked by anthropogenic sounds. These unheard voices might be warnings, finding of prey, or preparations of net-bubbling. When one species begins speaking louder, it will mask other species voices, causing the whole ecosystem to eventually speak louder.

According to the oceanographer Sylvia Earle, Undersea noise pollution is like the death of a thousand cuts. Each sound in itself may not be a matter of critical concern, but taken all together, the noise from shipping, seismic surveys, and military activity is creating a totally different environment than existed even 50 years ago. That high level of noise is bound to have a hard, sweeping impact on life in the sea.»

ADAPTATION AND MITIGATION

Much anthropogenic pollution ends up in the ocean. The 2011 edition of the United Nations Environment Programme Year Book identifies as the main emerging environmental issues the loss to the oceans of massive amounts of phosphorus, a valuable fertilizer needed to feed a growing global population», and the impact billions of pieces of plastic waste are having globally on the health of marine environments. Bjorn Jennssen (2003) notes in his article, "Anthropogenic pollution may reduce biodiversity and productivity of marine ecosystems, resulting in reduction and depletion of human marine food resources". There are two ways the overall level of this pollution can be mitigated: either the human population is reduced, or a way is found to reduce the ecological footprint left behind by the average human. If the second way is not adopted, then the first way may be imposed as world ecosystems falter.

The second way is for humans, individually, to pollute less. That requires social and political will, together with a shift in awareness so more people respect the environment and are less disposed to abuse it. At an operational level, regulations, and international government participation is needed. It is often very difficult to regulate marine pollution because pollution spreads over international barriers, thus making regulations hard to create as well as enforce.

Without appropriate awareness of marine pollution, the necessary global will to effectively address the issues may prove inadequate. Balanced information on the sources and harmful effects of marine pollution need to become part of general public awareness, and ongoing research is required to fully establish, and keep current, the scope of the issues. As expressed in Daoji and Dag's research, one of the reasons why environmental concern is lacking among the Chinese is because the public awareness is low and therefore should be targeted. Likewise, regulation, based upon such in-depth research should be employed. In California, such regulations have already been put in place to protect Californian coastal waters from agricultural runoff. This includes the California Water Code, as well as several voluntary programs. Similarly, in India, several tactics have been employed that help reduce marine pollution, however, they do not significantly target the problem. In Chennai, sewage has been dumped further into open waters. Due to the mass of waste being deposited, open-ocean is best for diluting, and dispersing pollutants, thus making them less harmful to marine ecosystems.

OIL POLLUTION TOXICITY TO MARINE FISH

Oil pollution toxicity to marine fish has been observed from oil spills such as the *Exxon Valdez* disaster, and from nonpoint sources, such as runoff, which is the largest source of oil pollution in marine waters. Crude oil entering waterways from spills or runoff contain polycyclic aromatic hydrocarbons (PAHs), the most toxic components of oil. The route of PAH uptake into fish depends on many environmental factors and the properties of the PAH. The common routes are ingestion, ventilation of the gills, and dermal uptake. Fish exposed to these PAHs exhibit an array of toxic effects including genetic damage, morphological deformities, altered growth and development, decreased body size, inhibited swimming abilities and mortality. The morphological deformities of PAH exposure, such as fin and jaw malformations, result in significantly reduced survival in fish due to the reduction of swimming and feeding abilities. While the exact mechanism of PAH toxicity is unknown, there are four proposed mechanisms. The difficulty in finding a specific toxic mechanism is largely due to the wide variety of PAH compounds with differing properties.

Research on the environmental impact of the petroleum industry began in earnest, during the mid to late 20th century, as the oil industry developed and expanded. Large scale transport of crude oil increased as a result of the increasing worldwide demand for oil, subsequently increasing the number of oil spills. Oil spills provided perfect opportunities for scientists to examine the in situ effects of crude oil exposure to marine ecosystems, and collaborative efforts between the National Oceanic and Atmospheric Administration (NOAA) and the United States Coast Guard resulted in improved response efforts and detailed research on oil pollution›s effects. The Exxon Valdez oil spill in 1989, and the Deepwater Horizon oil spill in 2010, both resulted in increased scientific knowledge on the specific effects of oil pollution toxicity to marine fish.

Exxon Valdez oil spill

Focused research on oil pollution toxicity to fish began in earnest in 1989, after the *Exxon Valdez* tanker struck a reef in Prince William Sound, Alaska and spilled approximately 11 million gallons of crude oil into the surrounding water. At the time, the Exxon Valdez oil spill was the largest in the history of the United States. There were many adverse ecological impacts of the spill including the loss of the loss of billions of Pacific herring and pink salmon eggs. Pacific herring were just beginning to spawn in late March when the spill occurred, resulting in nearly half of the populations eggs being exposed to crude oil.

Pacific herring spawn in the intertidal and subtidal zones, making the vulnerable eggs easily exposed to pollution.

Deepwater Horizon oil spill

After April 20, 2010, when an explosion on the *Deepwater Horizon* Macondo oil drilling platform triggered the largest oil spill in US history, another opportunity for oil toxicity research was presented. Approximately 171 million gallons of crude oil flowed from the seafloor into the Gulf of Mexico, exposing the majority of surrounding biota. The Deepwater Horizon oil spill also coincided directly with spawning window of various ecologically and commercially important fish species, including yellowfin and Atlantic bluefin tuna. The oil spill directly affected Atlantic bluefin tuna, as approximately 12% of larval tuna were located in oil contaminated waters, and Gulf of Mexico is the only known spawning grounds for the western population of bluefin tuna.

Exposure to oil

Oil spills, as well daily oil runoff from urbanized areas can lead to polycyclic aromatic hydrocarbon (PAHs) entering marine ecosystems. Once PAHs enter the marine environment, fish can be exposed to them via ingestion, ventilation of the gills, and dermal uptake. The major route of uptake will depend on the behavior of the species of fish and the physicochemical properties of the PAH of concern. Habitat can be a major deciding factor for the route of exposure. For example, demersal fish or fish that consume demersal fish are highly likely to ingest PAHs that have sorbed to the sediment, whereas fish that swim at the surface are at a higher risk for dermal exposure. Upon coming in contact with a PAH, bioavailability will effect how readily the PAH is taken up. The EPA identifies 16 major PAHs of concern and each of these PAHs has a different degree of bioavailability. For instance, PAHs with lower molecular weight are more bioavailable because they dissolve more readily in water and are therefore more bioavailable for fish within the water column. Similarly, hydrophilic PAHs are more bioavailable for uptake by fish. For this reason, usage of oil dispersants, like Corexit, to treat oil spills can increase the uptake of PAHs by increasing their solubility in water and making them more available for uptake via the gills. Once a PAH is taken up, the fish's metabolism can affect the duration and intensity of the exposure to target tissues. Fish are able to readily metabolize 99% of PAHs to a more hydrophilic metabolite through their hepato-bilary system. This allows for the excretion of PAHs. The rate of metabolism of PAHs will depend on the sex and size of the species. The ability to metabolize PAHs into a more hydrophillic form can prevent bioaccumulation and halt PAHs from being passed on to organisms further up the food web. Because oil can persist in the environment long after oil spills via sedimentation, demersal fish are likely to be continually exposed to PAHs many years after oil spills. This has been proven by looking at the biliary PAH metabolites of bottom dwelling fish. For instance, bottom dwelling fish still showed elevated levels of low molecular weight PAH metabolites 10 years after *Exxon Valdez* oil spill.

Crude oil Components

Crude oil is composed of more than 17,000 compounds. Among these 17,000 compounds are PAHs, which are considered the most toxic components of oil. PAHs are formed by pyrogenic and petrogenic processes. Petrogenic PAHs are formed by the elevated pressure of organic material. In contrast, pyrogenic PAHs are formed through the incomplete combustion of organic material. Crude oil naturally contains petrogenic PAHs and these PAH levels are increased significantly through the burning of oil which creates pyrogenic PAHs. The level of PAHs found in crude oil differs with the type of crude oil. For example, crude oil from the *Exxon Valdez* oil spill had PAH concentrations of 1.47%, while PAH concentrations from the North Sea have much lower PAH concentrations of 0.83%.

Sources of Crude oil Pollution

Crude oil contamination in marine ecosystems can lead to both pyrogenic and petrogenic PAHs entering these ecosystems. Petrogenic PAHs can enter waterways through oil seeps, major oil spills, creosote and fuel oil runoff from urban areas. Pyrogenic PAH sources consist of diesel soot tire rubber and coal dust. Although there are natural sources of PAHs such as volcanic activity and seepage of coal deposits, anthropogenic sources pose the most significant input of PAHs into the environment. These anthroprogenic sources include residential heating, asphalt production, coal gasification, and petroleum usage. Petrogenic PAH contamination is more common from crude oil spills such as *Exxon Valdez*, or oil seeps; however, with runoff pyrogenic PAHs can also be prevalent. Although major oil spills such as *Exxon Valdez* can introduce a large amount of crude oil to a localized area in a short time span, daily runoff comprises most of the oil pollution to marine ecosystems. Atmospheric deposition can also be a source of PAHs into marine ecosystems. The deposition of PAHs from the atmosphere into a water body is largely influenced by the gas-particle partitioning of the PAH.

Effects

Many effects of PAH exposure have been observed in marine fish. Specifically, studies have been conducted on the embryonic and larval fish, the development of fish exposed to PAHs, and uptake of PAHs by fish via various routes of exposure. One study on found that Pacific herring eggs exposed to conditions mimicking the ''Exxon Valdez'' oil spill resulted in premature hatching of eggs, reduced size as fish matured and significant teratogenic effects, including skeletal, cardiovascular, fin and yolk sac malformations. Yolk sac edema was responsible for the majority of herring larval mortality. The teratogenic malformations in the dorsal fin and spine, and in the jaw were observed to effectively decrease the survival of developing fish, through the impairing of swimming and feeding ability respectively. Feeding and prey avoidance via swimming are crucial for the survival of larval and juvenile fish. All effects observed in herring eggs in the study were consistent with effects observed in exposed fish eggs following the *Exxon Valdez* oil spill. Zebrafish embryos exposed to oil were observed to have severe teratogenic defects similar to those seen in herring embryos, including edema, cardiac dysfunction, and intracranial hemorrhages. In a study focused on the uptake of PAHs by fish, salmon embryos were exposed to crude oil in three various situations, including via effluent from oil coated gravel. PAH concentrations in embryos directly exposed to oil and those exposed to PAH effluent were not significantly different. PAH exposure was observed to lead to death, even when the PAHs were exposed to fish via effluent. From the results, it was determined that fish embryos near the *Exxon Valdez* spill in Prince William Sound that were not directly in contact with oil still may have accumulated lethal levels of PAHs. While many laboratory and natural studies have observed significant adverse effects of PAH exposure to fish, a lack of effects has also been observed for certain PAH compounds, which could be due to a lack of uptake during exposure to the compound.

Proposed Mechanism of Toxic Action

While it has been proven that different classes of PAHs act through distinct toxic mechanisms due to the variations in their molecular weight, ring arrangements, and water solubility properties, the specific mechanisms of PAH toxicity to fish and fish development are still unknown. The proposed mechanisms of toxicity of PAHs are toxicity through narcosis, interaction with the AhR pathway, alkyl phenanthrene toxicity, and additive toxicity by multiple mechanisms.

- The narcosis model was not able to accurately predict the outcome of PAH mixture exposure of herring and pink salmon, according to a study.
- The primary toxicity of these PAHs in fish embryos has been observed to be AhR independent, and their cardiac effects are not associated with AhR activation or Cytochrome P450, family 1, member A induction in the endocardium.
- The alkyl phenanthrene model has been studied by exposing herring and pink salmon to mixtures of PAHs in an attempt to better understand the toxicity mechanisms of PAHs. The model was found to generally predict the outcomes of sublethal and lethal exposures. Oxidative stress and effects on cardiovascular morphogenesis are proposed mechanisms for alkyl phenanthrene toxicity. The specific pathway is unknown.
- Since PAHs contain many different variations of PAHs, the toxicity may be explained by using multiple mechanisms of action.

REFERENCES

1. Administration, US Department of Commerce, National Oceanic and Atmospheric. "What is the biggest source of pollution in the ocean?". oceanservice.noaa.gov. Retrieved 2015-11-22.
2. Hamblin, Jacob Darwin (2008) Poison in the Well: Radioactive Waste in the Oceans at the Dawn of the Nuclear Age. Rutgers University Press. ISBN 978-0-8135-4220-1
3. Gerlach, S. A. (1975) Marine Pollution, Springer, Berlin
4. Young, Emma (2003). "Copper decimates coral reef spawning". Retrieved 26 August 2006.

5. Environmental Protection Agency. "Liquid Assets 2000: Americans Pay for Dirty Water". Retrieved 23 January 2007.
6. Washington State Department of Ecology. "Control of Toxic Chemicals in Puget Sound, Phase 2: Development of Simple Numerical Models", 2008
7. Panetta, LE (Chair) (2003) America's living oceans: charting a course for sea change [Electronic Version, CD] Pew Oceans Commission.
8. Farmer, Andrew (1997). Managing Environmental Pollution. Psychology Press. ISBN 0415145155
9. Podsadam, Janice (19 June 2001). "Lost Sea Cargo: Beach Bounty or Junk?". National Geographic News. Retrieved 8 April 2008.
10. Meinesz, A. (2003) Deep Sea Invasion: The Impact of Invasive Species PBS: NOVA. Retrieved 26 November 2009
11. Aquatic invasive species. A Guide to Least-Wanted Aquatic Organisms of the Pacific Northwest. 2001. University of Washington.
12. Pimentel, D.; Zuniga, R.; Morrison, D. (2005). "Update on the environmental and economic costs associated with alien-invasive species in the United States". Ecological Economics. 52 (3): 273–288. doi:10.1016/j.ecolecon.2004.10.002.
13. Coral Mortality and African Dust: Barbados Dust Record: 1965–1996 US Geological Survey. Retrieved 10 December 2009.
14. Duce, R. A.; Unni, C. K.; Ray, B. J.; Prospero, J. M.; Merrill, J. T. (1980). "Long-Range Atmospheric Transport of Soil Dust from Asia to the Tropical North Pacific: Temporal Variability". Science. 209 (4464): 1522–4. doi:10.1126/science.209.4464.1522. PMID 17745962.
15. Usinfo.state.gov. Study Says African Dust Affects Climate in U.S., Caribbean.. Retrieved 10 June 2007.
16. Prospero, J. M.; Nees, R. T. (1986). "Impact of the North African drought and El Niño on mineral dust in the Barbados trade winds". Nature. 320 (6064): 735–738. doi:10.1038/320735a0.
17. U. S. Geological Survey. Coral Mortality and African Dust.. Retrieved 10 June 2007.
18. Observations: Oceanic Climate Change and Sea Level In: Climate Change 2007: The Physical Science Basis. Contribution of Working Group I to the Fourth Assessment Report of the Intergovernmental Panel on Climate Change. (15MB).
19. Doney, S. C. (2006) "The Dangers of Ocean Acidification" Scientific American, March 2006.
20. Cheung, W.W.L., et al. (2009) "Redistribution of Fish Catch by Climate Change. A Summary of a New Scientific Analysis" Pew Ocean Science Series.
21. PACFA (2009) Fisheries and Aquaculture in a Changing Climate
22. Ahnert, A.; Borowski, C. (2000). "Environmental risk assessment of anthropogenic activity in the deep-sea". Journal of Aquatic Ecosystem Stress and Recovery. 7 (4): 299–315. doi:10.1023/A:1009963912171.

23. Halfar, J.; Fujita, R. M. (2007). "ECOLOGY: Danger of Deep-Sea Mining". Science. 316 (5827): 987. doi:10.1126/science.1138289. PMID 17510349.

24. Glasby, G. P. (2000). "ECONOMIC GEOLOGY: Lessons Learned from Deep-Sea Mining". Science. 289 (5479): 551–3. doi:10.1126/science.289.5479.551. PMID 17832066.

25. Sharma, R. (2005). "Deep-Sea Impact Experiments and their Future Requirements". Marine Georesources & Geotechnology. 23 (4): 331–338. doi:10.1080/10641190500446698.

26. Nath, B. N.; Sharma, R. (2000). "Environment and Deep-Sea Mining: A Perspective". Marine Georesources & Geotechnology. 18 (3): 285–294. doi:10.1080/10641190009353796.

27. Coral reefs around the world The Guardian, 2 September 2009.

28. Orr, J. C.; Fabry, V. J.; Aumont, O.; Bopp, L.; Doney, S. C.; Feely, R. A.; Gnanadesikan, A.; Gruber, N.; Ishida, A.; Joos, F.; Key, R. M.; Lindsay, K.; Maier-Reimer, E.; Matear, R.; Monfray, P.; Mouchet, A.; Najjar, R. G.; Plattner, G. K.; Rodgers, K. B.; Sabine, C. L.; Sarmiento, J. L.; Schlitzer, R.; Slater, R. D.; Totterdell, I. J.; Weirig, M. F.; Yamanaka, Y.; Yool, A. (2005). "Anthropogenic ocean acidification over the twenty-first century and its impact on calcifying organisms". Nature. 437 (7059): 681–686. doi:10.1038/nature04095. PMID 16193043.

29. Key, R. M.; Kozyr, A.; Sabine, C. L.; Lee, K.; Wanninkhof, R.; Bullister, J. L.; Feely, R. A.; Millero, F. J.; Mordy, C.; Peng, T. -H. (2004). "A global ocean carbon climatology: Results from Global Data Analysis Project (GLODAP)". Global Biogeochemical Cycles. 18 (4): n/a. Bibcode:2004GBioC..18.4031K. doi:10.1029/2004GB002247.

30. Raven, J. A. et al. (2005). Ocean acidification due to increasing atmospheric carbon dioxide. Royal Society, London, UK.

31. UNEP, FAO, IOC (2009) Blue Carbon. The role of healthy oceans in binding carbon

32. Monaco Declaration and Ocean Acidification A Summary for Policymakers from the Second Symposium on the Ocean in a High-CO2 World. Intergovernmental Oceanographic Commission of UNESCO, International Geosphere-Biosphere Programme, Marine Environment Laboratories (MEL) of the International Atomic Energy Agency, Scientific Committee on Oceanic Research. 2008.

33. Manado Ocean Declaration World Ocean Conference Ministerial/High Level Meeting. Manado, Indonesia, 11–14 May 2009.

34. Feely, R. A.; Sabine, C. L.; Hernandez-Ayon, J. M.; Ianson, D.; Hales, B. (2008). "Evidence for Upwelling of Corrosive "Acidified" Water onto the Continental Shelf". Science. 320 (5882): 1490–2. doi:10.1126/science.1155676. PMID 18497259.

35. ^ Jump up to: a b Milkov, A. V. (2004). "Global estimates of hydrate-bound gas in marine sediments: How much is really out there?". Earth-Science Reviews. 66 (3–4): 183–197. doi:10.1016/j.earscirev.2003.11.002.

36. The oceans occupy 361 million sq km
37. USGS World Energy Assessment Team, 2000. US Geological Survey world petroleum assessment 2000—description and results. USGS Digital Data Series DDS-60.
38. Selman, Mindy (2007) Eutrophication: An Overview of Status, Trends, Policies, and Strategies. World Resources Institute.
39. "The Gulf of Mexico Dead Zone and Red Tides". Retrieved 27 December 2006.
40. Duce, R. A.; Laroche, J.; Altieri, K.; Arrigo, K. R.; Baker, A. R.; Capone, D. G.; Cornell, S.; Dentener, F.; Galloway, J.; Ganeshram, R. S.; Geider, R. J.; Jickells, T.; Kuypers, M. M.; Langlois, R.; Liss, P. S.; Liu, S. M.; Middelburg, J. J.; Moore, C. M.; Nickovic, S.; Oschlies, A.; Pedersen, T.; Prospero, J.; Schlitzer, R.; Seitzinger, S.; Sorensen, L. L.; Uematsu, M.; Ulloa, O.; Voss, M.; Ward, B.; Zamora, L. (2008). "Impacts of Atmospheric Anthropogenic Nitrogen on the Open Ocean". Science. 320 (5878): 893–7. doi:10.1126/science.1150369. PMID 18487184.
41. Addressing the nitrogen cascade Eureka Alert, 2008.
42. Kroeger, Timm (2012). Dollars and Sense: Economic Benefits and Impacts from two Oyster Reef Restoration Projects in the Northern Gulf of Mexico. TNC Report.
43. Burkholder, JoAnn M. and Shumway, Sandra E. (2011). "Bivalve shellfish aquaculture and eutrophication". In: Shellfish Aquaculture and the Environment. Ed. Sandra E. Shumway. John Wiley & Sons.
44. Kaspar, H. F.; Gillespie, P. A.; Boyer, I. C.; MacKenzie, A. L. (1985). "Effects of mussel aquaculture on the nitrogen cycle and benthic communities in Kenepuru Sound, Marlborough Sounds, New Zealand". Marine Biology. 85 (2): 127–136. doi:10.1007/BF00397431.
45. Newell, R.I.E.; Cornwell, J.C.; Owens, M.S. (2002). "Influence of simulated bivalve biodepositon and microphytobenthos on sediment nitrogen dynamics, a laboratory study". Limnology & Oceanography. 47 (5): 1367–1379. doi:10.4319/lo.2002.47.5.1367.
46. Lindahl, O.; Hart, R.; Hernroth, B.; Kollberg, S.; Loo, L. O.; Olrog, L.; Rehnstam-Holm, A. S.; Svensson, J.; Svensson, S.; Syversen, U. (2005). "Improving marine water quality by mussel farming: A profitable solution for Swedish society". Ambio. 34 (2): 131–138. doi:10.1579/0044-7447-34.2.131. PMID 15865310.
47. Weisman, Alan (2007). The World Without Us. St. Martin's Thomas Dunne Books. ISBN 0-312-34729-4.
48. "Plastic Debris: from Rivers to Sea" (PDF). Algalita Marine Research Foundation. Retrieved 29 May 2008.
49. "Research | AMRF/ORV Alguita Research Projects" Algalita Marine Research Foundation. Macdonald Design. Retrieved 19 May 2009.
50. UNEP (2005) Marine Litter: An Analytical Overview
51. Six pack rings hazard to wildlife. helpwildlife.com

52. Louisiana Fisheries – Fact Sheets. seagrantfish.lsu.edu
53. "'Ghost fishing' killing seabirds". BBC News. 28 June 2007.
54. Weiss, Kenneth R. (2 August 2006). "Plague of Plastic Chokes the Seas". Los Angeles Times. Archived from the original on 25 March 2008. Retrieved 1 April 2008.
55. Venema, Vibeke (17 October 2014). "The Dutch boy mopping up a sea of plastic". BBC.
56. Moore, Charles (November 2003). "Across the Pacific Ocean, plastics, plastics, everywhere". Natural History. Archived from the original on 27 September 2007. Retrieved 5 April 2008.
57. Sheavly, SB & Register, KM (2007). "Marine debris and plastics: Environmental concerns, sources, impacts and solutions". Journal of Polymers & the Environment. 15 (4): 301–305. doi:10.1007/s10924-007-0074-3.
58. Weisman, Alan (Summer 2007). "Polymers Are Forever". Orion magazine. Retrieved 1 July 2008.
59. Thompson, R. C. (2004). "Lost at Sea: Where is All the Plastic?". Science. 304 (5672): 838. doi:10.1126/science.1094559. PMID 15131299.
60. Moore, C. J.; Moore, S. L.; Leecaster, M. K.; Weisberg, S. B. (2001). "A Comparison of Plastic and Plankton in the North Pacific Central Gyre". Marine Pollution Bulletin. 42 (12): 1297–300. doi:10.1016/S0025-326X(01)00114-X. PMID 11827116.
61. Moore, Charles (November 2003). "Across the Pacific Ocean, plastics, plastics, everywhere". Natural History Magazine.
62. Moore, Charles (2 October 2002). "Great Pacific Garbage Patch". Santa Barbara News-Press.
63. "Plastics and Marine Debris". Algalita Marine Research Foundation. 2006. Retrieved 1 July 2008.
64. "Midway's albatross population stable | Hawaii's Newspaper". The Honolulu Advertiser. 17 January 2005. Retrieved 20 May 2012.
65. Jordan, Chris (11 November 2009). "Midway: Message from the Gyre". Retrieved 13 November 2009.
66. "Q&A: Your Midway questions answered". BBC News. 28 March 2008. Retrieved 12 May 2010.
67. Rios, L. M.; Moore, C.; Jones, P. R. (2007). "Persistent organic pollutants carried by synthetic polymers in the ocean environment". Marine Pollution Bulletin. 54 (8): 1230–7. doi:10.1016/j.marpolbul.2007.03.022. PMID 17532349.
68. Tanabe, S.; Watanabe, M.; Minh, T. B.; Kunisue, T.; Nakanishi, S.; Ono, H.; Tanaka, H. (2004). "PCDDs, PCDFs, and Coplanar PCBs in Albatross from the North Pacific and Southern Oceans: Levels, Patterns, and Toxicological Implications". Environmental Science & Technology. 38 (2): 403–13. doi:10.1021/es034966x. PMID 14750714.

69. "Indigenous Peoples of the Russian North, Siberia and Far East: Nivkh" by Arctic Network for the Support of the Indigenous Peoples of the Russian Arctic
70. Grigg, R.W.; Kiwala, R.S. (1970). "Some ecological effects of discharged wastes on marine life". California Department of Fish and Game. 56: 145–155.
71. Stull, J. K. (1989). "Contaminants in Sediments Near a Major Marine Outfall: History, Effects, and Future". Proceedings OCEANS. 2. p. 481. doi:10.1109/OCEANS.1989.586780.
72. North, W. J.; James, D. E.; Jones, L. G. (1993). "History of kelp beds (Macrocystis) in Orange and San Diego Counties, California". Fourteenth International Seaweed Symposium. p. 277. doi:10.1007/978-94-011-1998-6_33. ISBN 978-94-010-4882-8.
73. Tegner, M. J.; Dayton, P. K.; Edwards, P. B.; Riser, K. L.; Chadwick, D. B.; Dean, T. A.; Deysher, L. (1995). "Effects of a large sewage spill on a kelp forest community: Catastrophe or disturbance?". Marine Environmental Research. 40 (2): 181–224. doi:10.1016/0141-1136(94)00008-D.
74. Carpenter, S. R.; Caraco, N. F.; Correll, D. L.; Howarth, R. W.; Sharpley, A. N.; Smith, V. H. (1998). "Nonpoint Pollution of Surface Waters with Phosphorus and Nitrogen". Ecological Applications. 8 (3): 559–568. doi:10.1890/1051-0761(1998)008[0559:NPOSWW]2.0.CO;2.
75. "What You Need to Know About Mercury in Fish and Shellfish". March 2004. Retrieved 19 May 2007.
76. Gollasch, Stephen (3 March 2006). "Ecology of Eriocheir sinensis".
77. Hui, C. (2005). "Mercury burdens in Chinese mitten crabs (Eriocheir sinensis) in three tributaries of southern San Francisco Bay, California, USA". Environmental Pollution. 133 (3): 481–7. doi:10.1016/j.envpol.2004.06.019. PMID 15519723.
78. Silvestre, F.; Trausch, G.; Péqueux, A.; Devos, P. (2004). "Uptake of cadmium through isolated perfused gills of the Chinese mitten crab, Eriocheir sinensis". Comparative Biochemistry and Physiology A. 137: 189–196. doi:10.1016/S1095-6433(03)00290-3.
79. Science News. "DDT treatment turns male fish into mothers." 5 February 2000. (By subscription only.)
80. Bocca, Riccardo (5 August 2005) Parla un boss: Così lo Stato pagava la 'ndrangheta per smaltire i rifiuti tossici. L'Espresso.
81. Chemical Weapon Time Bomb Ticks in the Baltic Sea Deutsche Welle, 1 February 2008.
82. Activities 2006: Overview Baltic Sea Environment Proceedings No. 112. Helsinki Commission.
83. Noise pollution Sea.org. Retrieved 24 October 2009.
84. Ross, (1993) On Ocean Underwater Ambient Noise. Institute of Acoustics Bulletin, St Albans, Herts, UK: Institute of Acoustics, 18.
85. Glossary Discovery of Sounds in the Sea. Retrieved 23 December 2009.

86. Fristrup, K. M.; Hatch, L. T.; Clark, C. W. (2003). "Variation in humpback whale (Megaptera novaeangliae) song length in relation to low-frequency sound broadcasts". The Journal of the Acoustical Society of America. 113 (6): 3411–24. doi:10.1121/1.1573637. PMID 12822811.
87. Effects of Sound on Marine Animals Discovery of Sounds in the Sea. Retrieved 23 December 2009.
88. Natural Resources Defense Council Press Release (1999) Sounding the Depths: Supertankers, Sonar, and the Rise of Undersea Noise, Executive Summary. New York, N.Y.: www.nrdc.org.
89. Fertilizer and plastic pollution are main emerging issues in 2011 UNEP Year Book, 17 February 2011. News Centre, United Nations Environment Programme, The Hague.
90. Jenssen, BM (2003). "Marine pollution: the future challenge is to link human and wildlife studies" (PDF). Environ Health Perspect. 111 (4): A198–A199. doi:10.1289/ehp.111-a198. PMC 1241462Freely accessible. PMID 12676633.
91. Kullenberg, G. (1999). "Approaches to addressing the problems of pollution of the marine environment: An overview". Ocean & Coastal Management. 42 (12): 999–1018. doi:10.1016/S0964-5691(99)00059-9.
92. Matthews, G. (1973). "Pollution of the oceans: An international problem?". Ocean Management. 1: 161–170. doi:10.1016/0302-184X(73)90010-3.
93. Warner R (2009) Protecting the oceans beyond national jurisdiction: strengthening the international law framework. Vol. 3 of Legal aspects of sustainable development, Brill, ISBN 978-90-04-17262-3.
94. Daoji L and Daler D (2004) "Ocean pollution from land-based sources: East China Sea, China" Ambio, 33: 1–2. Royal Swedish Academy of Sciences.
95. Plastic pollution". Encyclopaedia Britannica. Retrieved 1 August 2013.
96. Hammer, J; Kraak, MH; Parsons, JR (2012). "Plastics in the marine environment: the dark side of a modern gift". Reviews of environmental contamination and toxicology. 220: 1–44. doi:10.1007/978-1-4614-3414-6_1.
97. Hester, Ronald E.; Harrison, R. M. (editors) (2011). Marine Pollution and Human Health. Royal Society of Chemistry. pp. 84-85. ISBN 184973240X
98. Lytle, Claire Le Guern. "Plastic Pollution". Coastal Care. Retrieved 19 February 2015.
99. Nomadic, Global. "Turning rubbish into money – environmental innovation leads the way".
100. Mathieu-Denoncourt, Justine; Wallace, Sarah J.; de Solla, Shane R.; Langlois, Valerie S. (November 2014). "Plasticizer endocrine disruption: Highlighting developmental and reproductive effects in mammals and non-mammalian aquatic species". General and Comparative Endocrinology. doi:10.1016/j.ygcen.2014.11.003.

101. Barnes, D. K. A.; Galgani, F.; Thompson, R. C.; Barlaz, M. (14 June 2009). "Accumulation and fragmentation of plastic debris in global environments". Philosophical Transactions of the Royal Society B: Biological Sciences. 364 (1526): 1985–1998. doi:10.1098/rstb.2008.0205. PMC 2873009Freely accessible. PMID 19528051.

102. Driedger, Alexander G.J.; Dürr, Hans H.; Mitchell, Kristen; Van Cappellen, Philippe (March 2015). "Plastic debris in the Laurentian Great Lakes: A review". Journal of Great Lakes Research. 41 (1): 9–19. doi:10.1016/j.jglr.2014.12.020.

103. Chemical Society, American. "Plastics In Oceans Decompose, Release Hazardous Chemicals, Surprising New Study Says". Science Daily. Science Daily. Retrieved 15 March 2015.

104. Aggarwal,Poonam; (et al.) Interactive Environmental Education Book VIII. Pitambar Publishing. p. 86. ISBN 8120913736

105. Biello, David (June 5, 2011). "Are Biodegradeable Plastics Doing More Harm Than Good?". Scientific American. Retrieved 1 August 2013.

106. Eriksen, Marcus (December 10, 2014). "Plastic Pollution in the World's Oceans: More than 5 Trillion Plastic Pieces Weighing over 250,000 Tons Afloat at Sea". PLOS ONE. 9: e111913. doi:10.1371/journal.pone.0111913.

107. North, Emily J.; Halden, Rolf U. (1 January 2013). "Plastics and environmental health: the road ahead". Reviews on Environmental Health. 28 (1): 1–8. doi:10.1515/reveh-2012-0030.

108. "Marine Debris in the North Pacific A Summary of Existing Information and Identification of Data Gaps" (PDF). United States Environmental Protection Agency.

109. Daniel D. Chiras (2004). Environmental Science: Creating a Sustainable Future. Jones & Bartlett Learning. pp. 517-518. ISBN 0763735698

110. Knight 2012, p. 5.

111. Karleskint, George; (et al.) (2009).Introduction to Marine Biology. Cengage Learning. p. 536. ISBN 0495561975

112. Gregory, M. R. (14 June 2009). "Environmental implications of plastic debris in marine settings--entanglement, ingestion, smothering, hangers-on, hitch-hiking and alien invasions". Philosophical Transactions of the Royal Society B: Biological Sciences. 364 (1526): 2013–2025. doi:10.1098/rstb.2008.0265. PMC 2873013Freely accessible. PMID 19528053.

113. Parker, L. (2014). New Map Reveals Extent of Ocean Plastic. National Geographic

114. Fernandez-Armesto, F. (2006). Pathfinders: A Global History of Exploration

115. Moore, C. J. (2014) Choking the Oceans with Plastic. New York Times

116. Hill, Marquita K. (1997). Understanding Environmental Pollution. Cambridge University Press. p. 257. ISBN 1139486403

117. Rodríguez, A; et al. "High prevalence of parental delivery of plastic debris in Cory's shearwaters (Calonectris diomedea)". Marine Pollution Bulletin. 64: 2219–2223. doi:10.1016/j.marpolbul.2012.06.011.

118. Derraik, J. G. B. (2002) The Pollution of the Marine Environment by Plastic Debris: a review
119. Thompson, R. C.; Moore, C. J.; vom Saal, F. S.; Swan, S. H. (14 June 2009). "Plastics, the environment and human health: current consensus and future trends". Philosophical Transactions of the Royal Society B: Biological Sciences. 364 (1526): 2153–2166. doi:10.1098/rstb.2009.0053.
120. Brydson, J. A. (1999). Plastics Materials. Butterworth-Heinemann. pp. 103-104. ISBN 0750641320
121. (1973). Polyvinyl Chloride Liquor Bottles: Environmental Impact Statement. United States. Department of the Treasury (contributor).
122. Malkin, Bonnie (July 8, 2009). "Australian town bans bottled water". The Daily Telegraph. Retrieved 1 August 2013.
123. Susan Selke, Rafael Auras, Tuan Anh Nguyen, Edgar Castro Aguirre, Rijosh Cheruvathur, Yan Liu: Evaluation of Biodegradation-Promoting Additives for Plastics. In: Environmental Science & Technology. 49, 2015, p. 3769, doi:10.1021/es504258u.
124. Groff, Tricia (2010). "Bisphenol A: invisible pollution". Current Opinion in Pediatrics. 22 (4): 524–529. doi:10.1097/MOP.0b013e32833b03f8.
125. "EN 13432". Green Plastics.
126. Government of Canada (2014). "Oceans Act: Governance for sustainable marine ecosystems". Government of Canada. Government of Canada.
127. "AF&PA Releases Community Recycling Survey Results". Retrieved 3 February 2013.
128. "Life cycle of a plastic product". Americanchemistry.com. Retrieved 3 September 2012.
129. "Plastics". Epa.gov. Retrieved 3 September 2012.
130. (PDF) http://himachal.nic.in/WriteReadData/l892s/17_l892s/Circular_WATER%20BOTTLED-21798821.pdf. Retrieved 2 September 2016. Missing or empty |title= (help)
131. (PDF) http://www.nbsslup.in/assets/uploads/clinks/circular160607_-2.pdf. Retrieved 2 September 2016. Missing or empty |title= (help)
132. (PDF) http://aghp.cag.gov.in/pag/sites/pag/files/Avoiding%20use%20of%20bottled%20water%20during%20meetings,%20conferences%20etc..pdf. Retrieved 2 September 2016. Missing or empty |title= (help)
133. http://sikenvis.nic.in/ViewGeneralLatestNews.aspx?Id=4964&Year=2016. Retrieved 2 September 2016. Missing or empty |title= (help)
134. http://www.downtoearth.org.in/news/bihar-bans-plastic-packaged-water-bottles-48475. Retrieved 2 September 2016. Missing or empty |title= (help)
135. Staff Reporter. "Green rules of the National Games". The Hindu.
136. "National Games: Green Panel Recommends Ban on Plastic". The New Indian Express.

137. "Kochi a 'Museum City' Too | The New Indian Express | Page 426326". M.newindianexpress.com. 2016-02-08. Retrieved 2016-04-27.
138. "National Games 2015: Simple Steps To Keep Games Green - Trivandrum News - Yentha.com". yentha.com.
139. "Setting a New Precedent". The New Indian Express.
140. "The garbage patch territory turns into a new state - United Nations Educational, Scientific and Cultural Organization". unesco.org.
141. http://www.rivistasitiunesco.it/articolo.php?id_articolo=2073

Index